C000293740

HIGH HORSE RIDERLESS

by

L.T.C. ROLT

. . . that high horse riderless,
Though mounted in that saddle Homer rode

W.B. YEATS

with an introduction by
JOHN SEYMOUR

GREEN BOOKS

This edition published in 1988 by
GREEN BOOKS
Ford House, Hartland, Bideford, Devon EX39 6EE

First published in 1947
Reprinted with the kind permission of Sonia Rolt

Cover design by Thomas Keenes and Simon Willby;
Illustration from a drawing by Leo Davy
Typeset by Computype, Exeter

Printed by Robert Hartnoll (1985) Ltd
Victoria Square, Bodmin, Cornwall

British Library Cataloguing in Publication Data

Rolt, L.T.C. (Lionel Thomas Caswell),
1910-1974
High horse riderless.
1. Technological change. Social aspects.
I. Title
306'.46

ISBN 1-870098-14-5

CONTENTS

	Acknowledgements	vi
	Introduction by John Seymour	1
	Introduction	9
	PART ONE	
I	The Age of Faith	25
II	The Age of Reason	37
III	The Industrial Revolution	43
	PART TWO	
IV	The City at Work	57
V	The City at Home	64
VI	The City at Play	72
VII	The City Thinks	79
VIII	The Twilight of the Arts	86
IX	War	105
	PART THREE	
X	A Plea for the Individual	115
XI	Education for Freedom	121
XII	Work and Wealth	132
XIII	Social and Political Organisation	146
XIV	Art and Religion	155
	Conclusions	169

ACKNOWLEDGEMENTS

The publishers would like to thank the following for reproduction permission: Cambridge University Press for the extract from *The Wheelwright's Shop* by George Sturt, which appears in Chapter IV; A.P. Watt Ltd on behalf of Michael B. Yeats and Macmillan London Ltd for the stanza from the poem 'Nineteen Hundred and Nineteen' in *The Collected Poems of W.B. Yeats,* which is quoted at the opening of Part II, also for the prose extract from *The Cutting of an Agate* by W.B. Yeats, which is used in Chapter VIII; Faber and Faber Ltd for the poem 'Babel' from *The Collected Poems of Louis MacNeice,* which is quoted at the opening of Part III.

INTRODUCTION
by *JOHN SEYMOUR*

L.T.C. Rolt's great relevance to us today is that he was the first thinker – and remains practically the only one – to address himself to the problem of how we should use machinery sensibly. This problem has hardly been given any consideration although it is perhaps the most important immediately practical problem of our time.

Machinery is here – and we cannot uninvent it. It is not reasonable to expect two men to stand for twelve hours a day pit-sawing trees up into planks, with one down a pit with his eyes full of sawdust and the other perched on top of a log, when they both know perfectly well that what would take them weeks of the most boring labour could be done in a few minutes with a circular saw. There have been plenty of good writers – from Blake to D.H. Lawrence – who have condemned unreservedly the Industrial Revolution, attributing to it all the malfunctionings of our society. Lawrence in particular deplored the awful *ugliness* of the post-industrial world, and any person with two eyes in his or her head would agree with him wholeheartedly. But Lawrence never had to work in a saw-pit nor crawl about on his hands and knees wielding a pick-axe down a coal mine like his father did. The world is divided today between the great majority who welcome technology as the liberating force which is going to save us all from ever having to do anything hard or boring and make us all ecstatically happy, and the minority who, like Blake and Lawrence, condemn industrialism out of hand.

True, since Rolt's time, we have had a new approach to the subject in a concept developed principally by E.F. Schumacher: appropriate or intermediate technology. The early American colonists anticipated him, in the matter of ripping planks out of logs at least, by devising a simple multiple reciprocating saw driven by a water wheel. That, it seems to me, is a perfect example of appropriate technology. What Rolt would have thought of Schumacher's concept we cannot know: my belief is he would have welcomed it.

Rolt's especial position was moulded by two factors, as he describes very vividly in the autobiographical notes in this volume and also in his *Landscape with Machines* (reprinted in 1984 by Alan Sutton in the Sovereign series). The first factor was the extreme beauty and peacefulness of the countryside in which he spent his childhood: 'that part of the Welsh Marches which lies upon the borders of the counties of Radnor, Brecon, Monmouth and Hereford' (one wonders what Rolt would have thought if he had known that *every one* of these beautiful and ancient counties would be destroyed and lumped together by bureaucrats so that not one of them still exists!). This early experience gave him a standard by which he was to measure the world in future: he could not really be contented with a world that was so rapidly departing from that sylvan innocence. The second factor was his passion, which was never to leave him, for engineering and for machines.

Rolt could see the great *beauty* of machines, at least of machines that were the product of real craftsmanship and not just mass-produced artefacts banged together by unskilled labourers or even robots. And so he apprenticed himself to various engineering firms, agricultural engineers in Worcestershire, locomotive manufacturers in Stoke-on-Trent. He became thus a fully-fledged member of a dying breed: a real engineer, the sort of man who went about with the basic tools of his profession *that he had made himself* as part of his apprenticeship and who looked upon himself as a true craftsman. This sort of pride is not hubris. The owner of it is not proud of himself, but of his craft. He feels it a great honour to have qualified in such a demanding profession. He feels humility when he considers other masters of it, those people who had gone before and who had to do fine work with even fewer and more primitive resources, and humility at the realization that a lifetime is not long enough to master all there is to know.

But Rolt differed from most practitioners of a hard profession by never becoming too specialized to look around him all the while, to question the direction in which his life was leading him – and which way his profession was leading the world. As early as p.9 of this book he questions the tendency of people to over-specialize: 'As a result, probably never before in human history has so much knowledge been accompanied by so little wisdom.' How many professional engineers have those sorts of worries?

Rolt finished his apprenticeship before the Great Depression more or less put an end to engineering in England for the time being. Then his father's financial ruin (his father had been a prosperous *rentier,* more interested in shooting and fishing than in anything else) made it necessary for Rolt to go into partnership in a garage on the western outskirts of London. The garage specialized in what we now call vintage cars. Rolt retained a passion for these because they were fine pieces of well-designed machinery built by true craftsmen. However, he never came to love their mass-produced successors. He also began to notice the degradation of the countryside and of England's towns and villages, which was setting in *because* of the motor car. It was all getting far too far away from his vision of rural peace and beauty: the vision he had carried with him from his beloved border country. The fumes, the smells, the noise, the fearful accidents, the ugliness that the motor age was producing made it necessary for him to dissociate himself from the whole business. It was no longer a world in which humorous and eccentric young gentlemen or their fathers, or beautiful ladies in green hats drove about a lovely countryside on empty roads in finely built and tuned machines. It was a world in which great hordes of non-regarding people, who didn't know the difference between a cam shaft and a micrometer and who never looked at the beautiful countryside, just wanted to race from one ugly place to another in the shortest possible time in a vain attempt to escape from the awful boredom and meaninglessness of their lives.

He had to get away from this and he solved the problem by getting married and making his home aboard a converted narrow boat: one of the splendid wooden vessels, driven by great thumping diesel engines that any engineer could love, built to carry goods along the two thousand miles of England's narrow canal network. Thereafter he earned what little money he needed by writing, both journalism and books. Most of his books were written for enthusiasts in a variety of subjects and few of them were directed at the general reader. Thus, if you are a canal buff you will most certainly have heard of L.T.C. Rolt, and also if you are a railway buff, or interested in the lives of the great engineers. I came to Rolt when I started to cruise the canals. I found that, not only was he far and away the best writer on the canals of England, but there was much more to him than that. I found that underlying his canal writings there was a profoundly thought-out philosophy that illuminated everything the man had to say and which makes him unique and very important.

Quite by accident I came upon a novel he had written: *Winterstoke.* However one might judge it as a novel, it is an astonishing history of a microcosm of England and displays not only a very good grasp of real history (the history of ordinary people and not just a list of dates and the names of kings and battles) but also a deep understanding of what has happened in the last couple of thousand years. But it is in *High Horse Riderless* that Rolt really propounds his own unique philosophy. It is not in any way facile. It was fifty years before its time – he finished writing it in the early 1940s and his ideas will be acceptable to more people today than they would have been then, but it will still, alas, be many decades before a majority of thoughtful people are ready to accept them.

Central to his thought is the simple truth that the creative instinct in humankind finds satisfaction in the process of creation and not in the result. The modernist fails to realize this. H.G. Wells' Utopia is a creation of this failure. Rolt quotes twice Stephen Harding's great aphorism: *Laborare est Orare.* Work is true prayer, and he points out that in Harding's spirit work was not done for reward but for itself. This accords not only with the Christian doctrine but with Vedic doctrine too. The central message of Krishna to Arjuna at the battle between Good and Evil was that we must do right not because we think we will get a reward, or even succeed in what we are doing, but *because it is right.* Rolt examines what is known about thought and life in the Age of Faith with great penetration and sympathy. He points out that medieval scholars realized that humankind could not abolish evil for 'evil was one of the two poles between which mens' freedom operated'. Take away our freedom to choose evil and you take away our freedom.

Rolt wrote before we realized the full peril in which stands the life of this planet owing to the evil wrought by humanity, but he certainly anticipated it. In his discussion of the Celtic church in Britain, the pre-Augustinian church, he sees that there was no antipathy between Man and Nature. But later the Roman church, unfortunately, fell into an unnatural asceticism and began to deny 'Man's creaturehood' and to accept a doctrine that held that Man was an end in himself and that the rest of Nature was simply there to provide for his needs. The perfection of the human soul was all that mattered: how we treated our planet was irrelevant. Alas, we are still suffering from this delusion and our world is suffering bitterly with us. Rolt's analysis of the way in which the Church fell into intellectual pride as a result of this error is most penetrating.

His description of the transition from the Age of Faith to the Age of Reason is illuminating. He realizes that reason is a gift of God just as faith can be, and that we should use it, but he knows that if we use it wrongly it will mislead us, and he realizes that if we use reason we should also use faith. Reason alone can lead us out into the barren desert, and in Part Two of the book Rolt sees what we shall find there – what we have found in fact – the consequent spiritual desolation that has turned us into a destructive and self-destructive species. But, unlike most writers that nowadays the majority of moderns would call obscurantists, Rolt does not therefore decry or repudiate reason, just as he does not repudiate out of hand the machine. But, implicit rather than stated is the realization that only if humankind regains a spiritual dimension can we reconcile the apparently irreconcilable and become sane again.

Chapter III on the Industrial Revolution is perhaps the most important chapter in the book. He *knew* about the Industrial Revolution: he saw it very much as part of it and from the inside. He saw it in its power and glory – but he saw very clearly the hell to which it must lead. It was the 'high horse' of the title and was becoming an uncontrollable force on its own. It was subject to no bit or bridle, it rampaged mindlessly and furiously throughout the world and unless curbed would destroy it in the end. Rolt saw clearly that the doctrine of individualism would in fact produce its opposite. He thought back on the early monastic communities and the early medieval settlements at their best, and realized that the discipline of a true community can free the individual as no illusory 'freedom' can. *Laissez faire* means also *laissez mourrir. Laissez faire* led to, among other things, the Irish potato famine.

He sees, and I have never seen this view of things expressed before, that the Renaissance was 'the flower of an age past, not the root of one to be born'. When I read this it came as a revelation and I have been turning over the ramifications of the idea in my mind ever since. Although he acknowledges throughout the book that it was the Christian church that held western civilization together (and I feel, although I may be reading too much into it, that he thought it was the only thing that could save civilization now), he recognizes that the Church made great mistakes in trying to stifle scientific enquiry. When the Church wields too much temporal power it always misuses it, for the Church, founded upon whatever Rock it may be, is still an institution composed of people, and people are by definition imperfect. When the medieval church wielded too much power it stifled the creative instinct and its dissolution was inevitable. A truth that lies at the heart of Rolt's thinking is that the creative urge, like sexual love, can be perverted into the love of *power,* and it is this love of power that is the root of all human evil.

Rolt had studied Marx and recognized that Marx had certain insights into the workings of history, but he realized that marxism could never show the way towards a happy or holy society. When he wrote *High Horse Riderless* he could not have known the full extent of the failure of marxist philosophy to form the base of a successful society, but he certainly anticipated it. He anticipated also the dire effect that capitalist exploitation would have on the so-called undeveloped parts of the world; he realized that it would inevitably lead to the savage destruction of Third World natural resources, as unindustrialized nations, trying desperately to pay for the rubbish that the industrialized nations convince them they need, destroy their forests and rip up their countryside.

He saw too that the weakness of large-scale capitalism is that it does not seek to discover what peoples' real needs are and satisfy them. It seeks to discover instead what it can most easily produce, and then it creates desires for such products by massive and effective advertising. The planet is raped to provide more and yet more material to satisfy these apparently insatiable demands.

I certainly do not intend to write a précis of the book here, for I could never express Rolt's processes of thought and his conclusions as well as he does. All I can do is to urge the reader to read the book, and to read this book if she or he never reads another. Too much of our 'alternative' thinking comes from people who have never really done anything in the real world. The strength of Rolt is that he had done things. He had been part of and had helped to build the real world as it is now, and yet he saw its faults and contradictions, and saw where it was tending. In his Conclusions at the end of this book he tells us how to change things — how to bridle the 'high horse' and to turn its energies to good purposes instead of evil ones.

The final message of this book is one of cheer and hopefulness. Rolt accords to humankind a high and noble role and destiny: the Kingdom of God on Earth he claims: 'is not a misty theological chimera, it is a practicable goal within the range of our conscious ability to attain.'

This book is a trumpet call for us to strive to attain it, to be dismayed by nothing and to remember while we work that it is the process of creation that matters, not the result. If we get the process right we do not have to worry about the rest.

HIGH HORSE RIDERLESS

L.T.C. Rolt

INTRODUCTION

'An angel satyr walks these hills.'
Kilvert's Diary

THIS is an age of knowledge. We are inclined to forget that less than a century ago the majority of us were illiterate, and such knowledge as we possessed was largely inherited and intuitive. To-day radio, newspapers, magazines, pamphlets and books present us with a bewildering wealth of fact and theory upon every subject under the sun. The accumulated store of centuries of learning is yielded up for our delectation. We gorge ourselves on this knowledge indiscriminately and find the feast as indigestible as a surfeit of cream buns. To relieve the resulting mental bilious attack we then fly to the patent medicine of some ready-made and over-simplified ideology, resolving in the future to eschew rich diet and confine ourselves instead to such knowledge as directly concerns our particular sphere. In other words, we specialize. As a result, probably never before in human history has so much knowledge been accompanied by so little wisdom.

Following a statement such as this, a definition of terms becomes necessary. Knowledge, I would say, may be defined as the retention in the mind of facts and ideas, wisdom as the means whereby the individual sifts and values knowledge in relation to his personal experience of life. Unlike knowledge, wisdom cannot be instilled from without by education ; it is only possible to postulate conditions favourable to its growth. The essence of wisdom is philosophy, but such philosophy must be personal and closely related by the individual with his way of life. There is no wisdom in the uncritical and unrelated assumption of a ready-made philosophy laid down by others however admirable its principles may be. A mind absorbing knowledge without wisdom resembles an office without a filing system or, to pursue a different analogy, if knowledge be represented as a set of tools, wisdom becomes the ability to use them. Manifestly it is better to use simple tools expertly than to possess a bewildering assortment of complicated gadgets and either neglect or use them incompetently.

In this case what is true of the individual is true of the nation also. We believe, not without justification, that we have the knowledge and the materials with which to build a new and finer civilization, and while the old falls in ruins about our ears we vainly endeavour to agree upon an improved design. In this we find ourselves in the position of workers in a vast factory, each of whom plays a minute specialized part in the production of the finished article and is thus unable to comprehend the whole process by which the final result is achieved. The factory, however, produces results for good or ill because the basic principles to which the product must conform are agreed upon, and its resources are organized accordingly. To the task of building a better civilization no such fundamental principle has yet been determined which will both found and inform the whole complex structure. Intellectual specialist designers, because they lack practical workshop experience, produce blue-prints of new worlds which the workers cannot understand and which the Cost Department, the technicians and executives, reject. Conversely, in the light of their experience, each of these groups explains how their own particular part of the design should be carried out, detail plans which the harassed designer finds impossible to reconcile with his wider conception. Meanwhile, as this babel goes on, the complicated machinery runs on ungoverned, threatening to destroy not only itself but its operators.

Manifestly, the only solution of this impasse lies in the determination of a common principle deeper than all differences upon which to build ; a principle which will be acceptable to all, which can be understood by all, and for which all are prepared to make concessions. Here we can no longer pursue the simile of the factory where the specialist works for limited ends and quick results, for the acceptance of a common principle involves the integration on the part of every individual of his specialized knowledge with the common goal for which all must work and to which there can be no short-cut. In other words, nations, no less than individuals, must learn the function of their tools, must temper knowledge with wisdom.

The chapters which follow record the attempt of an average intellect to determine such a common principle and having done so to build a practical social framework upon the foundation of that principle.

The most enlightened and scholarly minds might with good

reason quail before such a project, so that I feel that, before setting forth upon what really amounts to a mental pilgrimage, the reader is entitled to some explanation as to why so formidable a task should ever have been undertaken by one so ill qualified to perform it.

In the first place, I would like to make it quite clear that the work was undertaken with humility and with full knowledge of my own lack of scholarship, not from any evangelistic notion or intellectual conceit, but as the result of a personal desire to resolve for myself questions which my own experience in England during the uneasy lull between two wars continually and with growing forcibility prompted. Being therefore so largely the fruit of experience, the chapters which follow might well have been cast in an autobiographical form; but I rejected this idea. For not only am I of the opinion that no one under forty is fitted to write autobiography, but I also felt that such personal history would constitute a sort of woolly clothing about an argument which I wished to make as clear and concise as possible. Furthermore, it is not always possible accurately to date the emergence of particular ideas, and it is therefore fatally easy to fit them to events which they in fact post-dated. There is a great deal of truth in Oscar Wilde's aphorism: "Man is least himself when he talks in his own person. Give him a mask and he will tell the truth."

This was the reasoning which led me to cast a personal testament in impersonal form, and the following brief autobiographical note, describing how it came to be written, must serve as my apologia and slender warranty.

My childhood during the first world war, and for a short while thereafter, was spent in that part of the Welsh Marches which lies upon the borders of the counties of Radnor, Brecon, Monmouth and Hereford. This is a region which has since become associated with the name of Kilvert, although it possesses a significance older and deeper as the inspiration of those great metaphysicals of the seventeenth century, Thomas Traherne, of Credenhill, and Henry Vaughan, of Crickhowell.

Before and below my home lay the valley of the Wye, here no longer a youthful mountain torrent but a staid adult, winding soberly through rich meadows by gravel spit and salmon pool. Only winter storms or melting snows occasionally roused her to a brown fury of spate and foam. This pastoral scene had for my childish eyes the comfortable familiarity of a well-known and friendly face, associated as it was with visits to the little market town by the river, and with rarer excursions farther afield to Hereford.

If this valley of the Wye seemed to me to be the gateway to England, then the two dark peaks of the Black Mountains, Rhiw Wen and Pen-y-Beacon, which overtopped the foothills behind the house, were the guardians of the gate of Wales. The sight of these two mountains, northward facing and therefore so often as dark and forbidding as their name, always held for me a feeling of awe and enchantment that only the elemental and timeless can generate. Not only did this frowning massif of old red sandstone seem primeval, but its people also. They still led the life of their remote ancestors, scarcely touched by time, a life so close to nature that when they rode past to market on their shaggy, sure-footed ponies, baskets at elbow and crying to their sheep, it seemed they brought the wildness of their mountain valleys with them.

I grew to know those valleys whose solitude the twin peaks guarded; the valley of the lost Priory of Craswall; lonely Olchon; the Vale of Ewyas hiding the jewel of ruined Llanthony where the Honddu, as though in remembrance, sings ceaseless plain chant. They were walled by the mountain ridges that stretched southward like long fingers from the knuckles of the northern scarp. I knew these high places, also; knew the tonic wind that set the cotton grasses dancing and blew the manes of the mountain ponies into their eyes as they stood rumps to windward, while cloud shadows leapt lightly the deep folds of the cwms. Often in the blue days of late August or early September I picked the purple-stained whinberry or the rarer scarlet cranberry and listened to the pipe of the curlew which seemed the very voice of the mountains—"their loneliness, the exultation of their stone." Below me I would see the small white farmsteads of a people who still spoke of spirits, yet who seemed to the stranger sullen and suspicious by reason of their intuitive resistance to the tide of modernism which beat ever more relentlessly against their mountain dykes. To-day these dykes leak and many of the higher farms stand empty and roofless, but then they still seemed secure, and the only rumour of another world was the glare of the distant open-throated furnaces of Ebbw Vale which, on clear nights, lit up the sky over Pen-y-Gader Fawr. Somehow this lurid glow became associated in my mind with a war which otherwise seemed remote and unreal, and was thus a satanic symbol of a world of violence and death somewhere beyond my mountains.

As yet the machine had hardly crossed the threshold of these valleys, and for the majority life still consisted of an arduous yet measured round of labour by hand and horse-flesh. In the mountains, the gleaming shears gathered the wool harvest and the flail thrashed out the meagre crops. On the richer farms of the vale I saw the great blue Hereford wagons with their well-groomed teams aglow with brass come swaying down the deep lanes at harvest home, leaving the ripe ears hanging like banners from the hazels.

In view of the foregoing it may seem strange that I should, at an early age, have developed a passion for things mechanical which made me resolve to become an engineer. To the best of my knowledge no member of my family has ever been connected with this profession, so my choice must remain unaccountable. Nevertheless my early surroundings undoubtedly influenced me deeply, how deeply I did not realise at the time, because I took them for granted. Their influence only became apparent later when they came into sharp contrast and conflict in my mind with the new mechanised world into which my profession led me.

Soon after we left the Marches and moved to a cottage on that part of the North Cotswolds that looks toward Bredon Hill and the Evesham Vale, I was sent to a public school where I spent what I shall always regard as the most unhappy and wasted years of my life. I never ceased to chafe against an inflexible curriculum which attempted to cram my head with unrelated facts, and against what I regarded as a criminal attempt to force the individual into a stereotyped mould instead of encouraging the particular growth of his individuality on creative lines. At the early age of sixteen I persuaded my parents to take me away from school so that I might begin a five years' engineering apprenticeship. The great day when I walked out of my school house for the last time remains evergreen in my memory, and I shall always regard it as the date when, with the exception of the three Rs, my education began.

The first two years of my apprenticeship were spent in the workshop of a large mechanised farm in the Vale of Evesham. This experience made a strong contrast with the traditional farming methods of my earlier recollection, but I accepted it without question as the result of a logical process of development. It was not until I subsequently embarked on a three years' term with a firm of locomotive engineers in the industrial Midlands that I first began to question the validity of a progressive modernism which I had hitherto accepted. For I found myself living between two worlds which, try as I might, I could not reconcile. During my brief holidays I would walk upon the close turf of the hills about whose feet the wide plain washed like a sea, rolling away to the Clees, the Malverns, and May Hill beyond the silver thread of Severn. For me this was the world of my childhood, a world of immutable reality, but now I could never put from my mind my other world of furnaces and tumultuous machines, nor could I ever forget the life of my workmates, an interminable monotony of days punctuated by the recurrent plodding to and fro along cinder tracks and rafty cobbled streets, between the isolated poles of the factory and the home; a home represented by the ranks of smoke-grimed tunnel-backs, in one of which I lodged.

It was at this time that I ran the gamut of modernist ideologies, believing that in them I should find the answer to the problem, and for a while I thought I had succeeded. For religion I had no use. My schooldays had given me a surfeit of religious moralism unrelated to life as I found it, and an overdose of pietism which I found equally indigestible. My reaction was to succumb to the popular worship of the human mind which I now perceive to have been no less than the "I think, therefore I am" of Descartes. I was strengthened in this belief by the sense of pride, amounting almost to arrogance, which I felt in my growing mastery over complex and powerful machines. When my hand held the regulator of a locomotive on the test track, or the controls of an electric travelling crane sweeping along under the roof girders of the clangorous boiler shop, I knew the exultation of power and felt convinced that the human mind, alone but unafraid in a mindless but predatory and therefore hostile universe, would eventually emerge as a triumphant conqueror, not only of the natural world but of its own defects. But here the contrast between my two worlds prompted the reflection that so far the advance toward this goal of ultimate perfectibility had not been conspicuously rapid. In fact to my eyes it appeared that mankind had retreated rather than advanced. Why?

I found, or rather I thought I found, the answer in orthodox socialism. The trouble was not fundamental but was simply due to defective and archaic economic and social machinery. Once the vested interests of monopoly capitalism with their fatal *laissez-*

faire philosophy had been overthrown and succeeded by a planned economy of production for need instead of for profit, all would be well, working hours would be reduced in proportion to the efficiency of the machine to afford leisure for all, the industrial worker for whom I had a profound admiration would come into his own, and the road to Utopia would be plain to see. And yet there was something in my childhood's recollection which raised faint doubts as to the validity of this argument, and which refused to be altogether suppressed no matter how hard I tried to dismiss it as mere sentimental nostalgia and defeatist reaction.

Nevertheless, these socialistic views were strengthened when, after two years, the locomotive works, though maintaining over a thousand men in full employment and with plenty of orders in hand, closed down for ever owing to the miscarriage of a piece of remote financial jugglery on the part of the chairman of directors. I was fortunate in being able to serve the remaining year of my apprenticeship elsewhere, but for the majority of my workmates this catastrophe marked the beginning of many vitiating and impoverished years on the dole.

The end of my full term of apprenticeship coincided with the great slump, and I found that now that I demanded more than an apprentice's wage, there existed no ready market for my skill in the industrial world. For a time I followed the bubble of temporary prosperity from firm to firm, but when this bubble burst, as it invariably did after a few months, the last to come were logically the first to go, and so I would find myself once more upon the march.

In the course of this industrial pilgrimage I made two disturbing and closely related discoveries. One was that individual skill and responsibility was a liability which was being eliminated from industry with bewildering speed, and the other that wages bore little or no relation to that skill, but were based upon rate of output. I found myself called upon to exercise all my skill and resource in a small agricultural engineering shop in the South for less than half the wages I earned for performing a purely mechanical, mindless and repetitive task in a large factory in the North. While engaged on the latter job I could not help recalling the workmates of my apprenticeship and trying to imagine their reaction to such a task. For despite the drabness and monotony of their lives they had still contrived to preserve a certain sturdy individuality and adaptability which was symbolised by their treasured tool chests filled with the emblems of their craft; velvet covered scrapers, keen as razors, forged and tempered with their own hands; oiled surface plates, wood encased; handmade calipers, verniers and scribing-blocks etched with their names. In a few years these men seemed to have become obsolete and I, though still in my twenties, felt obsolete too. It seemed to me that very soon the only work for skilled men would be the building of prototypes prior to mass production, or the tool-room work of making elaborate foolproof tools and fixtures for the ant-men of the new workshops. I made up my mind that I would cease to be a mobile slave at the whim of a system I could no longer tolerate and, with a partner, decided to open a small business of my own. In doing so it was not long before I discovered that the tide of modernism was all against the individualism of the small business with qualitative ideals and limited resources. It was then that I realised that the socialist views I had hitherto professed did not imply the restoration of those individual and qualitative values which I had grown to prize above all else, but their complete eclipse in a world of gigantic state planned and controlled monopolies beside which the so-called private monopolies of my experience were small fry. Whether these took the form of monopolies of specialized function or of cartels, controlling every stage of production from raw material to finished product, seemed to me to be irrelevant. In either event Socialism, by seeking to control the evil of individual acquisitiveness would destroy the good of individual ability.

This conviction had a profound effect upon my outlook, and I found that the natural world of my childhood once more assumed its place in my mind as the norm of reality, whereas the world of modernism appeared increasingly unreal and ephemeral in its tragic falsity, an abstraction built upon abstraction. Painfully I began to re-discover traditional and organic values and found that by applying them I could with increasing facility diagnose the maladies of "Progress". Yet, although I thought I could now perceive intuitively what was wrong, much still remained unresolved and puzzling to my mind. So far my criticism of modernism was almost wholly destructive and I could not feel that it was valid, or that it was any more than a form of defeatism unless I could formulate some practicable and more desirable alternative. Manifestly, to counsel the abandonment of all the scientific and technical discoveries of the Industrial Revolution, and a return to a machineless "simple life" was not only nostalgic but as impracticably Utopian as the Socialist dream of the Leisure State. As an engineer I felt certain that the machine in which I took a craftsman's interest was not in itself the villain of the piece, but that a fault, deeper than economics and organization, lay in the method of its application. What was that fault and how could it be rectified? How could mechanism and science be reconciled with the harmony which I perceived in the natural world and with the life of individual creative freedom which seemed to me to be the only social aim that was consonant with that harmony?

By this time the mechanical and urban world of modernism disquieted me profoundly, so dark was the shape of things to come which it seemed to me to portend. I became absorbed with the prospect of a return to the natural world which now seemed to me, not the romantic illusion of the modernists, but the only reality. But first I must solve my philosophic problem, to my own satisfaction at all events, otherwise such a return would be no forward step, but an escape. I felt that solution would be easier if only I could look objectively and with detached vision at the modern world from the standpoint of some "island" of reality set in the midst of it, and yet withdrawn from the headlong flight of "progress". This idea sounds a practical impossibility, but in 1939 it assumed concrete form in the shape of a canal boat. Not only was this literally an island, albeit a small one, but it was mobile, enabling me to come and go about England by my own secret ways, seeing but unseen. The story of one such journey I have already recounted in an earlier book. It might well have been a second ark that my wife and I had built, so soon after its completion did the material world around us dissolve in the chaos and ruin of the second world war.

Yet, despite the alarms of war and the fact that it compelled me to return to a profession I had decided to abandon, the idea was vindicated, and the boat became no Ivory Tower, but a firm base from which I found I could view the tragic world about me undismayed. More important still I began to find answers to my self-imposed problems; slowly, painfully, but surely, forging for myself a philosophy which no longer conflicted with my experience.

I found that the best way to shape my thoughts into concise and intelligible form was to write them down as though for the benefit of a general reader. The chapters which follow represent the result. It is said that there is nothing new under the sun, and I should be conceited indeed if I were to claim for them any original or revelatory character. They are merely the fruit of a deal of hard thinking. Yet I would ask the reader to believe that in a limited sense they are original because, as should now be evident, they are the product of experience rather than scholarship and therefore in no sense a compilation built out of a sequence of literary concepts. Obviously my ideas have been influenced by what I have read over a period of years, but life has been too short to be both practical man and scholar. I am fully conscious of my shortcomings in the latter respect and yet, if I were to have those years again I would not choose otherwise. Philosophies cannot be chosen from the peg like ready-made suits, and I think that a grain of wisdom acquired by experience is better than a whole harvest in a library. If that grain of wisdom is afterwards found to be corroborated by some great thinker of the past so much the better.

Because this book amounts to a rejection of materialism the reader may wonder at my apparent inconsistency in dealing so hardly with what I have called " Religious dogma." Some preliminary words of explanation on this point, which I consider a vital one, may not come amiss.

So soon as my views began to shape themselves into a philosophic pattern which was recognizably Christian by implication, I naturally became most concerned to discover the reason for the failure of that philosophy to retain its beneficent influence over the mind of western man from the Renaissance onwards. The birth of modern materialism has frequently been attributed to the influence of Hobbes and Descartes, and yet this explanation does not satisfy me. I regard the Cartesian philosophy as evil because it is arrogant, and yet I cannot believe that this evil could have scored such an overwhelming victory over the good in men unless the influence of the latter was already seriously weakened. In other words I doubt if the materialistic doctrine could ever have gained so rapid and so wide an acceptance if the mental climate of the time had not been readily disposed to receive it. Popular and successful philosophers are always those who swim with the mental current of their day. How then to account for this eclipse of Christendom?

The answer to this question must obviously remain for ever a subject for speculation, and, while it is one upon which I am ill-fitted to pronounce my view, it is my opinion, as I have attempted to explain in this book, that the Christian philosophy is founded upon the reflection of certain eternal and absolute truths, and that its failure was due, not to any inherent flaw, but to the lack of eloquent re-statement in the light of the new knowledge which was so rapidly spreading. In consequence an ever increasing number of persons began to evaluate knowledge, not in relation to Christian principles, but for its own sake as a potential source of power. Learning gave them arrogance when it should have taught them humility. At a crucial period of history when men were beginning to shape their lives by reason rather than by intuitive faith, the eternal verities became a dead dogma instead of a living doctrine. A doctrine, in the definition of my dictionary, means that which is taught, whereas a dogma is defined as assertion without explanation or proof, and to my mind there is a world of difference between instruction and assertion. An age of Faith may have been satisfied with a dogma, but the age of Reason required a doctrine, and because no such re-valuation of the Christian Philosophy was forthcoming, it embraced the doctrine of materialism.

Let me give one illustration of the distinction which I draw between doctrine and dogma. In Chapter X of this book I have endeavoured to develop the theory that the root of evil and the cause of our modern ills is attributable to the frustration of the creative instinct in man, which results in its perversion into the will-to-power or into that masochism which is simply the reverse of the same medal. This theory is one of the foundation stones upon which this book has been built.

A friend of mine pointed out with perfect truth that this theory is simply a paraphrase of St. Augustine's: "What could begin this evil will but pride that is the beginning of all sin?" Yet with all humility I would submit that the latter statement, despite its unquestionable truth, has to-day become a dogma, a bare assertion no longer potent to influence the mind of modern man towards good. He may be prepared to acknowledge its truth but he will be powerless to apply it since he cannot perceive its relation to contemporary life and thought.

In our pre-industrial and predominantly rural past the life of man was largely governed by inherited principles which he accepted tacitly, and in which he believed intuitively without necessarily being able to ascribe reasons for his belief other than the warranty of tradition. This not only applied to the great problems of human existence, but to every detail of his life at home and in the fields. He acted in this way and in that way because his forefathers had done so since time immemorial. In brief, his wisdom exceeded his knowledge, he used simple tools but he used them with inherited and expert skill. For example, the history of craftsmanship or of any individual craftsman presents the same story of a tradition of skill handed down from father to son, and to which each generation made his individual contribution. The craftsman, in fact, is a living testimony to the validity of the theory of the inheritance of acquired characteristics. The modernist, however, impatiently dismisses tradition as reactionary and "unprogressive," without pausing to consider the profound significance of its implication. For the wisdom of the craftsman is not merely his own, it is the cumulative experience of many generations. This slow but organic process of building up wisdom and skill, generation by generation, constitutes true progress and, so long as its links remain unbroken, human society will continue to develop and expand upon a stable basis by a process of continual renewal and enrichment. If, however, the links are arbitrarily broken, then each successive generation inherits no birthright; any wisdom and skill the individual may acquire during his lifetime is solely the product of his own initiative and, in all probability, dies with him. This is the process which has occurred in the modern world and its result is the disintegration of society. The only remedies the modernist can suggest are improved education, which is of itself no remedy, or Utopian pleas for longer life such as Shaw's *Back to Methuselah*.

In the past two hundred years traditional wisdom has been discarded in the arrogance of newly-won knowledge which went to men's heads like immature wine. Yet I believe that to-day we are approaching, by the painful and bitter road of experience, the next and most momentous phase in which we are destined to discover that this wisdom which our fathers brusquely dismissed was actually based upon a more realistic philosophy than our own; that we shall learn to accept with faith and with reason what was once accepted with faith alone.

Already, through the medium of our more enlightened scientists, we are being made aware of the disastrous effects of our wholesale exploitation of natural resources for the sake of cash returns. While the outlook for the future seems dark I prefer to hope that we shall go further until we re-discover, in the light of clear reason, that the Christian philosophy of moral law constitutes the only sound basis of human contentment and social stability.

I have not only been concerned to express this belief, but to attempt to indicate the possible shape of its practical application in the sphere of social organization. I am of the opinion that modernism with its glib, never to be fulfilled, promises and plans of Leisure States and new Utopias must first be defeated on its own ground, and that it is not enough to reveal its fallacies and to counsel a change of heart. Because many people in this country are deeply disquieted in mind and are beginning to question things hitherto regarded as beyond question, I believe that, in a limited sense, this change of heart already exists. Yet for the lack of any constructive alternative, the majority are prepared to follow the lead of the collectivists along the road which leads to the servile state.

This book records the quest of a mind in search of such an alternative. It is not a blue print of a " brave new world ", but merely an attempt to re-discover the road which, rough and ill-signposted though it might be, Western Christendom once followed. In the course of my search I have blundered heavy-footed into many fields which we regard to-day as the preserves of the specialist. I have no doubt that I shall be warned off and told that I should have been better advised to remain among my machines. Yet I am quite unrepentant. I have already declared my belief that we have become over-specialized, not only physically but mentally, and that in the search for truth which we all instinctively pursue we should seek to widen rather than restrict our terms of reference. It is only by so doing that we can enlarge our experience and our understanding. Nor is this all. By learning more and more about less and less, the specialist becomes increasingly exclusive and arrogant. His special knowledge spells power. It is his mystery. Though he wears no black cloak and steeple hat he is the twentieth century magician. If, on the other hand, we seek to learn a little about more and more, the better do we perceive the beauty and intricate order of the world around us. The older we grow the more clearly do we realise how little we know, and so our knowledge brings to us, not pride but humility, and with humility the sense of wonder.

PART I

. . . . . . But in these cases
We still have judgment here; that we but teach
Bloody instructions, which, being taught, return
To plague th' inventor: this even-handed justice
Commends th' ingredients of our poison'd chalice
To our own lips . . . . .
. . . . . . I have no spur
To prick the sides of my intent, but only
Vaulting ambition, which o'erleaps itself,
And falls on th' others.

—*Macbeth*. Act I, Scene 7.

PART ONE

Chapter I

THE AGE OF FAITH

The recorded story of man's existence on this planet and of his never-ending efforts to come to terms with his environment is filled with paradox and seeming contradiction, a flickering kaleidoscopic pattern of light and darkness which represents no less than the eternal war of good with evil motives which is implicit in his individual nature. This complexity is not rendered simpler when we realise that any record of the past other than a mere catalogue of facts and dates must inevitably reflect the prejudice of the human, and therefore far from omniscient, historian. A completely impartial judgment of past events is impossible; their relative importance is assessed by the judgment of the historian which is in turn affected by the prevailing trend of contemporary thought. Perhaps the nearest we can approach to impartiality is to survey the prevailing tendencies and influences of a period and ignore the exceptions which are said to prove rules. By so doing we may be guilty of over simplification, but will avoid the danger of losing sight of the wood in the dense undergrowth of wars and political intrigue.

The task of the twentieth century historian is rendered more difficult by the immense gulf of the Industrial Revolution which, more far-reaching in its effects than any the world has yet seen, divides him from the period he seeks to portray. Straining across this gulf, distance and the mists of the Renaissance distort our vision of the dim shape of mediæval England. Thus the romantic traditionalist glimpses forms of barbaric splendour and pageantry, of chivalry and faith, the nostalgic chimera of sentimentality. The so-called " progressivist ", on the other hand, sees only a darkness of tyranny, superstition and inhuman cruelty, and even goes so far as to employ the term mediæval to describe the Nazi " new order ". Both these views present a one-sided exaggeration, for good and evil cohabit in every age. Of the two, that of the " progressivist " is the more myopic, for to compare the Nazi order with an age which gave us " Piers Plowman " and an architecture of unmatched splendour is to display not merely gross ignorance but wilful blindness.

For the Age of Faith has left for our judgment its deathless monuments in stone, of which perhaps the greatest are the Cistercian houses of Rievaulx, Jervaulx, Tintern and Fountains. No matter how unimaginative and materialistic he may be, who can stand alone in the stillness of these ruins without experiencing some feeling of wonder, awe and, perhaps, humility at the passionate austerity and breadth of their conception? What manner of men were these who built to such high purpose? What brought about their downfall, and why should that purpose which could produce so glorious a flowering have perished from the earth?

These are the questions we ask ourselves while the jackdaws wheel overhead accentuating the silence with their harsh crying, and to imagination's ear there sounds the pad of sandalled feet, or from ruined choir the echo of the chant.

So soon as we are back on the road, once narrow, miry and thronged with pilgrims, but now a tarmac highway, our self-esteem is quickly restored. Familiar objects, petrol pumps, telegraph poles and hurrying cars look friendly and reassuring, for it is as though the past, like a prosecuting counsel, had for a moment come to life and pointed an accusing finger. We console ourselves with the popular impression that these abbeys, like the pyramids, were built by slaves; slaves ruled with superstition by a priesthood whose piety was a cloak for fat living. We picture the monk, fat, sly and rubicund, tippling rare wines while the beggar starves at the gate. Thus the modern world clouds our vision and we do not endeavour sincerely to answer the questions which the past has prompted.

Christianity was the motive force which laid the foundations of a new civilization amid the chaos and dark ruin of Europe after the fall of the Roman Empire, and the Cistercian order founded by Stephen Harding and St. Bernard constitutes what was perhaps the noblest attempt ever made to live in accordance with the basic principles of the Christian faith. The ruins they have left to us as witness are thus not merely the symbol of a religious sect, but of a way of life. The great church was the central symbol of faith about which all the manifold activities of a self-supporting community revolved. The Cistercian lay-brother was neither a slave nor an anchorite, but a skilled craftsman who wrought in metal, wood and stone, who built roads, wove cloth, bred stock and planted trees, and who tilled the soil of field and garden to make barren wastes fruitful. Yet all these manifold and highly individualistic activities were undertaken, not for personal enrichment, but for the benefit of the community and as an article of a faith which was summed up in the precept of Stephen Harding: "Laborare est Orare."

If the reader should imagine that this picture of mediæval life is a distortion or misrepresentation of fact he would do well to study one of those village histories which are the fruit of much painstaking research on the part of some local antiquary into Manorial and other early records. One such history, that of the Worcestershire village of Tardebigge[1] whose manor of Hewell was once in the demesne of the Cistercian Abbey of Bordesley, contains some illuminating references to mediæval activities and organization, certain of which are worthy of quotation here. For example, we obtain a glimpse of the varied activities of the Cistercian community from the brief account of the case, recorded in the Pleas of the Royal Forest of Feckenham when, in 1280, two Bordesley lay brothers were caught shooting (with what success we are not told) at the King's deer. "John of the Tilery (de la Teylereye) and John, son of William le Stodhurde (studman?), both probably servants of the abbey, were shooting at the King's deer in the forest, and were arrested and brought back to the Abbey. As they approached certain of the brethren came running from the swine farm (porcaria) and the shearing house (cissoria) to beg off the delinquents. Eventually the forester accepted half a mark and released them."

Again, we obtain an excellent idea of the diversity of Cistercian husbandry from the annual return of the bailiff of the demesne lands for the year 1459-1460. From approximately 200 acres of arable the following harvest was delivered to John Wycte, "co-monk and granator", i.e. the monk in charge of the granary: 94 quarters, 7 bushels, 3 pecks of wheat; 72 quarters, 1 bushel, 1 peck of barley; 30 quarters, 1 bushel, 1 peck of pulse; and 61 quarters, 7 bushels of oats. It is also interesting to note that out of an additional 7 quarters, 7 bushels of malt to be disposed of, 5 quarters, 2 bushels were used "for the expenses of the autumn" which presumably means ale for the harvesters. We are not told the acreage of pasture, but they supported three bulls, forty oxen, fifty-nine cows, twenty-six steers and heifers, fourteen bullocks and

[1] *"A Thousand Years in Tardebigge"*—Margaret Dickinson.

yearlings, forty-three calves, two boars, five sows, thirty-four pigs, forty-five small pigs and ten sucking pigs. It is curious that sheep, which contributed so greatly to the downfall of the monasteries, are not mentioned.

But the Cistercians were not only farmers on their own account, they were also landlords, and in order to gather some idea of village life and organization under their stewardship we must glance at the records of the Manorial Courts. In the particular history under consideration court records are not quoted until after the dissolution when, in 1534, the Abbey surrendered to Henry VIII, but as in this instance the village organization was destined to continue substantially unchanged, it is not irrelevant to consider the later records given. From them we learn that so far from being the abject serfs of an autocratic petty dictator as is so often popularly supposed, the villagers, free and copy holders, governed themselves. They possessed a delicate and highly organized system of government which was, in the most literal sense, " government of the people, by the people, and for the people ", and which makes our modern conception of democratic government appear to be the political abstraction which in fact it is. The basis of this government was " the Custom of the Manor ", a system developed and perfected by many generations of local inhabitants to suit local conditions. Its laws were thus recognised as a product of cumulative wisdom and they were accepted " according to custom from time immemorial ". Obviously this custom varied in detail from region to region, but the principle remains the same so that a single example may be taken as typical.

It would be impossible, within the compass of one short chapter, to examine in detail and in all its significance this Manorial Custom which was the essential scaffolding of village life. It is of immeasurably greater importance historically than the records of battle and the rivalries of princes and powers with which school history books are filled, and for this reason a general outline must be given, the relevance of which will become apparent in later chapters.

The basic principle of Manorial Custom was the law of Frankpledge which was established before the Norman Conquest. This laid down that the inhabitants of a district were responsible for any crime committed by one of their number, in other words the village community were themselves responsible for keeping the peace and maintaining law and order within their own boundaries. Frankpledge was thus the essence of self-government as opposed to a form of control from without such as our modern forms of government represent. In the case of the particular example under consideration, Custom appointed two courts, the Small Court, or Court Baron, which met every two to three weeks, and the Great Court, Court Leet or View of Frankpledge, which met twice a year, in spring and autumn, and whose chief business was the election of officials. In considering the duties of these officials it is essential to remember that each tenant under the open field system possessed a share or strip in the common fields. He possessed such a share in each field so that no individual should monopolize the best land. The size of these shares or strips was determined by a unit of measurement based on work done ; thus in the arable fields the unit was a selion, the amount ploughable by a single ox team in one day, and in the meadows a math or one day's mowing by one man. In addition the tenant had the right to graze his beasts on the common or waste, but no more than he could maintain through the winter. It was the duty of the officials appointed by the Court to see that these and other privileges were not abused, and that hedges, timber and roads were properly cared for and maintained. Firstly, then, the bailiff of the Lord of the Manor chose the Foreman of the Court Jury, while the Reeve or collector of rents chose the second man. These two jurymen were then responsible for the choice of fourteen to sixteen of their fellow tenants to make up the Jury or " Verdictors ." This Jury then elected the officers. In this particular instance where a large and scattered parish was involved the district was divided into " Tythings " for each of which an elected " Tythingman " or Constable was responsible. Other officers were: Mowers (in charge of meadows), Haywards (in charge of hedges), Woodwards, Searchers of Leather and Tasters of Bread and Ale, these latter being appointed to report upon quality and prices and ensure that the " Assize " was maintained. This " Assize " was no less than the mediæval law of the just price, and to " break the Assize " was to sell goods which were too dear or of inferior quality. The Jury also appointed two Affeerers whose duty it was to consider the justice of the fines or "Pains" which the Jury imposed.

All tenants were bound to attend court under penalty of fine, unless specifically excused, and it was the duty of the official concerned to " present " his case before the Jury. For example, in the records of the Tardebigge Court several shoemakers are presented by the Searcher of Leather and fined because they " sold shoes and took excessive gain". Similarly the tasters presented "William Lyll" because he "sold ale before it was tasted, and therefore the taster does not know how many times he brewed". Other fines were inflicted for overstocking the common pasture, for trespass of beasts in the corn, for keeping hedges and gates unclosed and for cutting down trees without reference to the Woodward. Fines are also recorded against persons who attempted to encroach upon the common lands, and the building of cottages without an allocation of at least four acres of land was forbidden. Obviously a self-supporting community could not countenance an unregulated expansion of its population and for this reason we also find that sub-tenants and lodgers or "inmates", as they were called, were discouraged.

That the jury of this village court dispensed justice without fear or favour we learn when we read of the occasion, in the seventeenth century when they successfully "presented" the Lord of the Manor for stopping an ancient right of way, and how on another occasion several members of the jury fined themselves. They also defended themselves stoutly when, after the Reformation, centralized authority first began to impose its will upon them. "In 1595", we read, "Jocosa Sheward found a horse and colt doing damage on her land and impounded them" (in the village pound where such strays were kept and subsequently sold if not claimed within a year and a day). "Henry Penforde released them by order of the Sheriff of Worcester, and the Homage said that this ought to have been done by order of the steward of the manor and fined Henry Penforde 2d." The following year the jury presented that William and Richard Lewes had broken the liberties of this manor by apprehending the bodies of John Lewes and others on an order of the sheriff, without licence of the lady of this manor, of the steward, or of the bailiff, or of the bedel, "to the prejudice of the jurisdiction of this court and in contempt of the lady of the manor". And later, "Henry Cottrell was fined 6s. 8d. because he had impleaded Richard Lewes in the chancery of the Lady Queen at Westminster for customary lands held of this manor which ought to be argued within this manor. And he was ordered to cease from the prosecution of the suit in chancery and seek remedy as is lawful in this court under pain of 40s." These events may seem trivial to be quoted at length, but this is far from the case, because they are concrete examples of the value which our "primitive" village forefathers set upon their rights and liberties of self-government and individual responsibility, rights so soon to be filched away by that new tide in the affairs of men which the Reformation ushered in.

These two institutions, the Cistercian community and the open-field village are, in the light of subsequent events, by far the most significant features of the mediæval period. The former was a magnificent experiment in the art of living in strict accordance with Christian principles which developed and decayed within the confines of that period, while the latter was an established system which came to its fullest flower under monastic stewardship, but which both pre-dated and survived it. It survived until it was destroyed from without by hostile and predatory forces of the emergence of which the decay of the monastic system which led to the Reformation was the first symptom.

The tragic story of the corruption and consequent destruction of the Cistercians is as important as the bright chapter of their success. In fact their tragedy was largely the result of that success. Their good husbandry was so well rewarded that the rule of material poverty and spiritual wealth was reversed, labour ceased to be an end, it became a means; a means to acquire wealth and therefore power. Following the Black Death, which caused an acute labour shortage, a considerable proportion of monastic land hitherto devoted to mixed farming was converted into sheep-walks. The Abbeys specialized, reaping great profit as wool merchants, so much so that the secular interest of the crown feared their power and coveted their gains with the result that their dissolution quickly followed.

Precisely how corrupt the monasteries had become at the period of the Reformation must remain open to question, for in considering contemporary records the fact that those responsible for the annexation would naturally exaggerate its justification must be taken into account. Nevertheless, the downfall of the old church which marked the end of the age of faith was not merely a material matter of immoderate wealth on the one hand and cupidity on the other; these were simply the outward manifestation of more fundamental changes which call for wider consideration.

The Christian church first realised the power of its influence over the minds of men when, at its behest, a united Europe rose and with common purpose marched across two continents to the relief of the Holy Sepulchre. Never before or since has Europe been so united by a single faith and purpose, so that it must have seemed to many that the conception of a united Christendom would become an accomplished fact. Yet it was the very magnitude of this response to its call which was destined to bring about the church's downfall. The First Crusade made the church aware that it wielded a weapon of tremendous physical force and it proceeded to abuse it. There followed a period of ecclesiastical domination, the axis of which was Rome. Rome maintained a rigid orthodoxy by force, claiming not only temporal supremacy, but infallibility for the Papacy. Any who dared oppose it were considered heretics, to be punished with ruthless ferocity, and the crusaders became the army of a dominant power, a punitive weapon of coercion.

As the power of the church waxed and its wealth increased it ceased to be an effective advocate for the great principles of which it was the guardian, because it no longer acted in accordance with them. Humility and poverty gave way to arrogance and wealth. At a critical time when men's minds were athirst for knowledge, and seeking to recover the great legacy of Greece, the church failed to supply the leaven of wisdom. Instead, as it became secularized, what had once been a living doctrine became, for lack of re-statement, an inflexible dogma. Assertion took the place of instruction, and on this account the church sought jealously to retain a monopoly of knowledge and to punish as heretics those who upheld the freedom of the intellect. The result was a priesthood as dogmatic and bigoted as that which Jesus had so strenuously opposed in his lifetime and which finally contrived his death.

In view of the foregoing it may be wondered why Christianity survived as a unifying force so long as it did. The answer is that although the heirarchical tree of the church may have been withering at the top, its roots went deep and were still sound. The lack of physical communications in the Middle Ages made it impossible for any contemporary to grasp the pattern of current events and so approach to anything like the historian's synoptic vision. Unless its consequences chanced to touch him nearly the average man remained in ignorance of the subtleties of Papal policy and of the crimes perpetrated in the name of the Faith. For him the church remained the symbol of his faith and the ultimate arbiter of right and wrong. In the exercise of this function the church did not consist of an institution imposed upon the community to enforce certain rules of conduct with threats of hell fire. It was not so much the source as the expression of belief, a belief not confined to its four walls on one day of the week, but informing every aspect of the life and work of the village community in a manner so intimate and inseparable as to be almost beyond the conception of the modern mind. A wealth of custom and ceremony which the Christian church inherited from an earlier paganism gave point to this intimacy by illustrating and acknowledging man's dependance upon the eternal mysteries of the natural world for his daily bread, and so sanctifying the work of the fields. They were thus the grace and crown of labour. The heritage of imaginative and vigorous carvings which the mediæval village masons have left to us reveal more clearly than any words the depths of this intuitive belief. The winged cherubim is no more out of place in the farm house than is the corbel head of some long dead yeoman in the church. Was not the founder of the faith a carpenter, the Christ of the Trades, and did he not express himself in the language of the field and the workshop?

This organic and symbolical trinity of church, field and home was destined to survive for many years until it was finally sundered by hostile external forces. But meanwhile the seeds of those forces were already germinating in the form of a reaction against the church's power and immoderate wealth. A new growth of scepticism and disaffection took root among the kings and nobles directly involved in Papal intrigues.

Among the first of these secular figures destined to question the power of the priesthood was the Emperor Frederick II of Germany. He it was who embarked on the farcical sixth crusade having belatedly committed himself to undertake the venture in return for Papal support of his claim to the throne. In the course of this expedition he effectually stripped the last shreds of glamour from the crusading spirit by concluding, in the best traditions of power politics, a bloodless "gentleman's agreement" with the Sultan of Egypt. He subsequently incurred Papal displeasure, but remained throughout his life a liberal-minded sceptic, evincing a humanistic outlook readily understandable to-day because it was the precursor of modernism. His bland indifference to the excommunications and thunderous interdicts of successive Popes, coupled with his telling denunciation of the pride and corruption of the higher priesthood could not fail to exercise a profound influence over the minds of his powerful secular contemporaries and successors.

It was this new spirit of scepticism and doubt in high places which was fated to bring about the Reformation and sound the death knell of the Age of Faith. Its impetus was accelerated by the wider dissemination of knowledge and ideas outside the confines of the church which the invention of the printing press made possible.

Yet the intellectual climate which provoked an event of such momentous and far-reaching importance in human history as the Reformation was so complex, that to diagnose it in such simple terms would be a gross simplification. Scepticism was only one of its facets. Another element, also closely associated with the spread of knowledge, and of equal importance, was that change in the attitude of man toward the natural world which we call Puritanism.

The early church, particularly the early British church which was of Celtic foundation, recognized no antipathy between man and nature. Despite the mysticism of the Celtic saints and the magnificent grandeur of their philosophic concepts, they lived so close to natural reality that they never lost that humility which recognizes man's creaturehood, nor that sense of wonder which perceives, in the order and beauty of the natural world, the handwriting of a creator. From this vision sprang the conception of a natural world which demonstrated a divine order, and of man as a part of that order. But owing to the perilous gift of knowledge which caused man's fall, man's will was free, a freedom which placed upon him a great responsibility, because it meant that for him alone of created things participation in that order was not obligatory but deliberate. The early Christian philosophers recognized this, making it their task to interpret the laws which governed the natural order, and to apply them to human society. This was the principle underlying the whole body of mediæval Canon Law. Recognizing the essential imperfections of mankind, these mediæval schoolmen harboured no Utopian illusions of human perfectibility, and realized that, at its best, human society could only be a pale reflection of that natural order upon which they sought to model it. They could not abolish evil because evil was one of the two poles between which man's freedom operated, but so long as the conception of the natural order remained clear in men's minds it provided a standard by which such evil could unfailingly be measured and so recognized. It was the function of the church, not only to expound this doctrine by its teaching, but to symbolize and celebrate the natural order and its creator.

The church of Rome, on the other hand, partly through contact at an early stage of its development with certain degenerate nature cults, and partly as a consequence of the wealth and power which it attained, began to develop a tendency to reject the natural world as the source of vanity and illusion. This attitude crystallized in extreme form in such sects as the Gnostics, the Monophysites and the Manicheans, and may be summarized as Catholic Puritanism. It was frequently manifest in a fanatical asceticism which aimed at the renunciation of " the world, the flesh and the devil ", and was therefore essentially masochistic. Unfortunately for these ascetics, however, the more strenuously they strove to exorcise the world and the flesh, the more obstreperous became the devil. The reason for this phenomenon was that by rejecting the former, which is in reality the recognition and affirmation of man's essential creaturehood, they were unconsciously exalting man and therefore the unique faculties by which man is distinguished. By thus extolling mind to the exclusion of matter, and by thus regarding nature as evil, man becomes evil " not "—to quote St. Augustine—" because that is evil to which it (the mind) turns, but because the turning itself is perverse ". Furthermore, the rejection of the world of nature as vanity and temptation leads readily to an absorption in human knowledge and the worship of man. By rejecting the world man forgets his creaturehood to become arrogant and proud so that the masochism of the fanatical religious ascetic readily falls a prey to the sadism of the will-to-power and the idolatry of man.

It was this strain of Catholic Puritanism which, strange and paradoxical though it may seem, contributed to the wealth, corruption and secularization of the late mediæval church besides being the ancestor of the great Puritan revolution which accompanied and followed the Reformation. Both regarded the world as a vale of illusion and travail through which it was ordained that man, the soldier of God, must fight his way; his success in that fight being the measure of his reward or punishment hereafter. This view was utterly alien to that of the early church which had sought to interpret the Jus Naturale, and consequently its great doctrines ceased to live. The eternal principles upon which Canon Law had been based were clouded, so that its enforcement became a matter of custom rather than conviction. Stephen Harding's great phrase " Laborare est Orare " was essentially realistic, and was not concerned with any question of reward, in fact it meant the precise opposite. It implied that by creative labour man was fulfilling his true function in the divine order, and that by his contribution he not only affirmed the existence of that order, but its creator also. But to the Puritan the phrase meant something very different. To him labour meant the fight of the soldier of God through a hostile world, a fight in which the end justified the means. The end was reward hereafter, and if the means happened also to bring earthly reward in the shape of riches, these also were the measure of virtue and proved the strength of righteousness. That by this process some acquired great wealth, while others sank to abject poverty was easily explained. The poor were those for whom God had ordained the greater travail, or who were lacking in their ability to "fight the good fight". While Christian charity should be shown to them, their condition was not due to any fault in the structure of society, but to the dispensations of an inscrutable providence.

The importance of this change in religious thought cannot be over-estimated. Unless it is properly appreciated it will be impossible to understand how men contrived, in the centuries after the Reformation, to reconcile their immorality in public affairs with private religious scruples which were still strong. In the next chapter I have given a brief sketch of that period in which I have endeavoured to trace the spread of what I have called the Machiavellian doctrine. This was born of the sceptical attitude towards the rule of the church which has already been mentioned, but it could not have so readily taken root had not religious thought itself undergone this vital change. Similarly the philosophy expounded by Descartes and his successors, which could never have been propounded, much less entertained in the mediæval world, found fertile soil for its arrogant yet desolate assumptions. By such means the whole intellectual climate was changed, and in the process the truth of St. Augustine's precept was proved. For by rejecting the world man became more worldly, and by turning from nature as evil he became evil because he had rejected the only standard by which evil could be measured.

In England the Reformation was exemplified by the house of Tudor. In our history books the Middle Ages went down with the banner of the Sun in Splendour at Bosworth Field, but a victory for Plantagenet would only have postponed those great changes in the order of society which had become inevitable.

Chapter II

THE AGE OF REASON

The changes in mental outlook, not only among the secular powers but within the church also, made the retirement of the church from its position of dominance in the affairs of Europe inevitable. Not only did the mediæval principles of the Jus Naturale fall on a world in which their meaning was no longer fully understood and to which they were therefore no longer applicable, but the church which sought to apply them lacked the force of precept and example because it had lost the vision which inspired them. For a time this church, jealous of its power, attempted to retain it by securing a monopoly of knowledge, but such a process of repression could not long continue, and the Reformation saw the flood of new learning breach these ecclesiastical dykes.

The result was such a release of creative power that the period has become known in history as the Renaissance. In Italy Da Vinci, Michael Angelo, Titian, Tintoretto and Paul Veronese, together constituted what was perhaps the greatest constellation of genius in the visual arts which the world had seen since the age of Pericles, while in England there arose the literary immortals Spenser, Shakespeare, Bacon, Marlowe, Sir Thomas Overbury and Ben Jonson. Because of the lustre of this genius, the Renaissance is often regarded erroneously as a golden age, whereas upon reflection it becomes obvious that such pinnacles of achievement can never be climbed in a single generation. Genius of this order is inevitably a plant of slow growth, the flower of generations of slowly accumulated learning, tradition, and creative power. In the hothouse of the Renaissance the carefully tended mediæval buds soon blossomed and as soon became over-blown. The artistic glories of the late sixteenth and early seventeenth centuries, in fact, represent the final achievement of the Middle Ages, and the expression of the new age is not to be found in the work of Shakespeare or Ben Jonson, but in the voice of Niccolo Marchiavelli.

The seeds of independence and scepticism which the intolerant dogmatism of the church had sown inspired the Renaissance nobleman with a new, cynical and profoundly materialistic ideology of which Machiavelli was not so much the author as the acute observer and recorder.

In the Middle Ages all men, rich and poor alike, were equal in the eyes of the church in so far as each, according to his degree, possessed rights and responsibilities in accordance with the moral law which the church laid down. Machiavelli, on the other hand, believed that a ruler by right of birth was an exalted being of more than common stature, and one whose conduct could not therefore be judged by the standard governing lesser men; that the acquisition and exercise of power was the profession of the truly great, the ultimate aim of human endeavour, and that any unscrupulous methods of cunning, duplicity and ruthlessness were justified in the attainment of this end. In the light of future events it is impossible to exaggerate the importance of this perverted and cynical philosophy of temporal absolutism, and its emergence is by far the most significant feature of the Renaissance period. By substituting the abstraction of temporal expediency for the realism of the mediæval conception of good and evil, it snapped the moral yardstick of human conduct which had hitherto formed the standard of human society, and which enshrined the hope of a united Christendom. By representing the conduct of government in human affairs as a battle of wits without rules, by thus fixing a gulf between governors and governed and breaking the ties of mutual responsibility between them, the Machiavellians at one stroke set at naught the Christian goal of the brotherhood of man. At first confined to a small heirarchy of princes and nobility, its influence accompanied the growth of commercialism and knowledge, filtering downwards through the strata of society while the spiritual discipline of the church receded before it. In so doing it separated rich from poor, employer from employed and strong from weak, to create a mental climate of which the mechanistic philosophy of Descartes was the logical and consequently the inevitable expression. This process produced two of history's many paradoxes, first that the power of monarchy was destroyed by its own philosophy, and second that freedom was lost in the pursuit of individual "liberty". Both can best be illustrated by a brief survey of English history following the Reformation.

Henry VIII, who was the instrument of the Reformation in England, was a typical example of the Renaissance prince.

Whereas his predecessors had acknowledged the spiritual dominion of the church and had hesitated to incur its censure, Henry did not hesitate to defy Papal authority and stoutly maintained his "divine" right to rule over both church and state within the borders of his own realm. By this move the monarchy secured unquestioned freedom of action, but its supremacy was to be short-lived since it involved individual responsibility which could not escape the consequences of those actions. Though the arrogant Tudors escaped this unavoidable Nemesis which awaits unlimited power, its blow was to fall later on the hapless house of Stuart.

A large proportion of the monastic lands forfeited at the dissolution passed into the possession of the new aristocracy which had grown up about the Tudor court, following the eclipse of the Plantagenet nobility. These men, courtiers, counsellors and merchants, proved themselves to be apt pupils of their Sovereign by asserting that might was right. To make their sheepwalks and parks, land held by peasant and yeoman from church or crown, by right of custom since Saxon days, was seized and enclosed. They built themselves great houses such as Compton Wynyates or Westwood which, beautiful as they are, symbolize, by their withdrawal from the village, the gulf which was opening between Lords and Commons. They were the forerunners of the alien Palladian mansions of the eighteenth century, screened from the vulgar gaze by their high park walls. While the bellies of the Commoners remained empty the new aristocracy amassed great riches, for it was becoming increasingly difficult, in this world of expanding commerce, to apply the mediæval principle of the just price. The sharp edge of the sword of Canon Law, which, in the hands of the church, had proved so deadly a weapon against the activities of the usurer and the engrosser, soon became blunted by temporal expediency when wielded by secular authority.

The success of the Wool Lords was accompanied by the rise to power of the middle-man, in this case the so-called "master-weaver", such as the celebrated "Jack of Newbury", whose great weaving sheds, employing gangs of the new landless labour, were the first ancestors of our modern factory system.

Although this process of enclosure and exploitation only directly involved a fifth of rural England, it was the significant portent of the shape of things to come being no less than the practical application of Machiavellian principles. Expressed in the monetary terms of an increasingly commercialized world, these principles became the merchant's conception of freedom; freedom of acquisition and absolute ownership, freedom "to do what one would with one's own", to buy in the lowest and sell in the highest market, the freedom of the individual at the expense of the community. Obviously this glorification of selfish ambition and self-interest, masquerading as "liberty" or "self-determination", is the precise opposite of that which it claims to be since it results in the exploitation of the poor by the rich, of the weak by the strong. It explains the apparent paradox that every "reform" carried out in the name of this "liberty", from the period of the Renaissance down to the present day, has actually resulted in a loss of liberty for the majority.

In the consideration of the historical events of the seventeenth and eighteenth centuries in England it is essential to realize that this new commercial autocracy found its corporate voice as the Government of England, and in so doing became conscious of its immense power. Only one stumbling block stood in its path and that was the monarchy. Its conception of freedom could never be fully realized so long as the king retained his supreme right to command rich and poor alike. Claiming to represent the will of the people, Parliament brought increasing pressure to bear upon the king. Charles Stuart, however, refused to abrogate his authority, stubbornly maintained his "divine right", and the famous struggle between monarchy and constitution ensued. Although the Commonwealth was shortlived, and monarchy soon restored, the power of that monarchy had fallen with the head of Charles I.

Power over both church and state, having brought about the downfall of monarchical rule, now devolved upon a group which claimed to represent the people while it continued to act in the King's name. The great responsibilities which such power incurs were thus to a large extent dissipated, so that although such a government made individual tyranny impossible, it enabled a ruling caste of astute and powerful individuals to manipulate the political machine in the name of the state for their own ends and yet avoid retribution. In international affairs the "state" behaved like Machiavelli's prince without incurring his risks, and in so doing founded a tradition of diplomatic cauistry which survives in modern power politics. Meanwhile at home the whole principle of government underwent a profound change, which led to the gradual overthrow of that body of Canon Law which secular authority had inherited from the mediæval schoolmen. Professor R. H. Tawney[1] acutely analysed the nature of this change when he wrote: "The process by which natural justice, imperfectly embodied in positive law, was replaced by positive law which might or might not be the expression of natural justice, had its analogy in the rejection by social theory of the whole conception of an objective standard of economic equity. The law of nature had been invoked by mediæval writers as a moral restraint upon economic self interest. By the seventeenth century a significant revolution had taken place. 'Nature' had come to connote, not divine ordinance, but human appetites, and natural rights were invoked by the individualism of the age as a reason why self-interest should be given free play."

Perhaps the most significant and far-reaching enactment of this new regime was the Tonnage Act of 1694, whereby the control of money issue passed from the Crown to the Bank of England, whose original "created" loan of £1,200,000 to the Treasury of William III "upon a fund of perpetual interest" first established the national debt. From this small beginning there grew up the gigantic hydra-headed monopoly of banking and credit, the "funding system" against which William Cobbett was later to inveigh. Usury, condemned by the schoolmen as a venial sin, became the basis of trade.

The new philosophy and the new economics brought about the rise to power and prosperity of the urban commercial classes, the "people of the middle sort", while, to quote Thorold Rogers,[2] "the English people who lived by wages were sinking lower and lower, and fast taking their place, in contrast with the opulence which trade and commerce began, and manufacturing activity multiplied, as the beggarly hewers and drawers of prosperous and progressive England". A new type of landlord, who looked upon land as capital, changed the whole conception of land tenure, a change which virtually eliminated the small freeholder by favouring the accumulation of land in a few hands, and which had the effect of increasing the rents of arable land ninefold in the space of a century. New taxes were a heavy burden, while owing to the activities of those who bought to sell again, prices rose out of all proportion to wage increases.

The effect of these changes upon the Commons of England, who had hitherto lived on and by the land, was disastrous, and many

[1] *"Religion and the Rise of Capitalism."* [2] *"Six Centuries of Work and Wages."*

were compelled, in order to eke out a slender livelihood, to take up some form of trade, at first in addition to, but later to the exclusion of their husbandry. The Trade Guilds, which had hitherto sought to regulate entry into trade in proportion to local demand, found themselves powerless to control an expansion driven by the force of necessity. England's industrial areas were thus founded in the first instance, not by any process of migration from country to town, but by a great intensification of domestic industry in those rural areas where the necessary materials were readily available. Large scale commercial enterprise in our sense of the term was very largely confined to the basic industries which supplied the small producer with raw material, and to the companies of merchant venturers who marketed his finished products. Because the output of these products was no longer regulated by local needs, the activities of the latter perforce became increasingly aggressive, expansive, and ruinously competitive.

Despite the privations and abuses which this unregulated expansion of trade inevitably generated, the impoverished countryman still clung to his slender stake in the land and laboured in his own workshop, so that he still enjoyed a measure of independence. It was not until the latter half of the eighteenth century that the technical discoveries which ushered in the Industrial Revolution enabled the forces of capital to invade the realm of manufacture with the result that the Factory System silenced the domestic workshop forever, and the erstwhile Commoner of England became a dependant, owning nothing but his skill.

Chapter III

THE INDUSTRIAL REVOLUTION

The freedom of the intellect which was won at the period of the Renaissance gave to mankind a wealth of achievement in art, science and philosophy, the value of which is assured and immortal. Yet because the spirit of the new age was fundamentally materialistic, and obsessed with relative rather than absolute values, newly-won knowledge was used, not for the betterment of mankind as was frequently claimed, but to satisfy individual ambition in the pursuit of the power which wealth represented. The growing mercantile middle class became the patrons of science in their eagerness to exploit fresh fields of profit. Harnessed to the principle of maximum profit, scientific invention laboured to "improve" the process of manufacture by reducing the labour required to produce a given article so that it could be made in greater quantity and at less cost. Though this aim might adversely affect quality, the merchant's profit was assured, and the process might enable him to secure at least a temporary monopoly.

This powerful combination of scientific knowledge and commercial opportunism brought about the greatest revolution in the recorded history of mankind. In the brief interval of a century the whole material aspect of man's life and work which had hitherto been subject to a process of slow evolution over a period of thousands of years was suddenly and drastically transformed. It is extremely difficult for us to weigh the significance of the effects of this tremendous change because it is still proceeding on an accelerating scale, and we are therefore denied the historian's objective view. Nevertheless, it behoves us to examine and criticize as dispassionately as we can the road we are following, rather than to accept the easy and all too popular assumption that it represents inevitable progress which must be taken for granted and into the frame of which the individual life must be fitted.

Though the materialism favourable to the revolution was already a powerful force in seventeenth century England, it accomplished little until the second half of the next. This was due to two closely related causes, the lack of motive power and of an adequate transport system. England was still an association of self-supporting rural communities, and the minority of urban population depended on the products of the country in their immediate vicinity and upon water-borne trade. Without her market gardens and her river London could not exist, and these factors controlled her growth. Manufacturing capacity was dependent on man and horse-power aided to a very limited extent by harnessing wind and water. A great concentration of labour would thus have been required in order to manufacture on a large scale, and this was not possible since transport could neither keep it supplied with food and raw material, nor distribute the finished product.

Both these difficulties were overcome when Newcomen built, and Watt improved, the steam engine, while James Brindley constructed the Worsley to Manchester canal for the Duke of Bridgewater. The latter made possible the first industrial concentration, and the story of the growth of South Lancashire is typical of that which took place in other areas.

It is recorded that in 1757 the combined population of Manchester and Salford was 20,000, or approximately that of a fair sized market town to-day. 2,400 families were engaged in the production of textiles from wool by means of the spinning wheel and the hand loom. The trade was literally a family one, the master with his children and apprentices working and eating together. Samuel Smiles,[1] commenting on the Lancashire weaver of this period, says: "He was part chapman, part weaver and part merchant, working hard, living frugally, principally on oatmeal, and contriving to save a little money. As trade increased, his operations became more subdivided, and special classes and ranks began to spring into importance." This reference to the growth of specialization and the class system along with the expansion of trade has for us a significance undreamt of by Smiles.

Meanwhile Manchester had not yet severed its links with rural life and tradition. The district was interspersed with pastures, orchards and cornfields, one of which occupied the site of St. Anne's Square as late as 1770. Salmon was plentiful in the Medlock, and trout in its tributaries Irk and Tib. Wedding processions from the country accompanied by fiddlers frequented the "Ring o' Bells" and the "Blackamore," inns which were soon

[1] *"Lives of the Engineers,"* Vol. I. (Population figures quoted from the same source.)

destined to be demolished under the so-called Hunts Bank improvement scheme.

In 1761 the canal from the Worsley collieries was completed, and in 1773 its extension to the Mersey at Runcorn linked Manchester and Liverpool. The essential pre-requisite of industrial expansion having thus been provided, the revolution was fairly launched, and in a brief thirty-five years between 1770 and 1804 Manchester was transformed. In 1774 the population was 41,032, in 1801 84,020, and in 1821 187,031. For his great share in this expansion Francis Egerton, Duke of Bridgewater, was hailed by writers of the period and for long after as a great "benefactor". It is therefore pertinent to note that he paid his engineer Brindley half-a-crown a day for his work, while he received during this period an income of £80,000 a year in canal revenue alone.

Coincident with this opening up of communications and no less important was the revolution in the application of power to manufacturing processes, which became possible when James Watt developed Newcomen's crude non-rotative pumping engine into a rotative engine suitable for driving mill machinery. The steam engine made practicable the merchant's dream of larger turnover and greater profit, and its effect upon industry was immediate. In the case of Manchester the change was represented by the spinning jenney and the power loom which made obsolete the traditional hand tools of the old family business. The more powerful weaver became a mill owner, while his weaker competitor lost his business and became an employee so that the "special classes and ranks" remarked by Smiles quickly emerged.

Machinery thus provided the ideal weapon with which the rich and powerful could exploit their Machiavellian conception of freedom, and they proceeded to make the fullest use of it. South Lancashire became the most monstrous testament of the result of human greed that the world had yet seen. In the valleys of the Pennines there sprouted the tall chimney stacks of the new mills, while around them, like a malignant growth, crept, alley by alley and row by row, the wretched tenements of the mill workers. Here spawned under conditions of almost unbelievable squalor the new poor of the Age of Reason. The rivers became open sewers, fields where corn had waved a few years previously were blackened forever, while a perpetual pall of smoke darkened the sky.

Incredible as it may seem, the manufacturer, as the merchant had now become, still contrived to pose as a public benefactor. He accomplished this by employing at a starvation wage those whose livelihood he had robbed, and by supplying his goods over a wide area to the ruin of local industry in that area. His production, in fact, was in no way related to needs, but was simply carried on for maximum profit. Thus there arose the necessity of an expanding market in order to dispose of the goods made, and an increased flow of fuel, food, and raw material, in order to produce them. Water transport became inadequate to meet the demand, but the coming of the railway era in 1825 provided the answer, and inaugurated a new cycle of expansion the magnitude of which may be judged from the fact that by 1861 the population of Manchester had swollen to 460,028. Elsewhere industrial expansion was equally rapid, and the more powerful manufacturers began to appear in public life and to question the right of the great land-owners to govern England. The materialistic misconception of freedom which had led monarchy to overthrow the church, and aristocracy the monarchy now brought about the downfall of the aristocrat by the merchant. Just as the church paid lip-service to temporal power and monarchy had become a hollow figure-head of empty splendour, so aristocracy was destined either to become the pliant and pretentious but irresponsible tool of money power or perish.

The issue of cheaper food on which the industrialist went to battle against the landlord was well chosen and plausible. In his rôle of public benefactor the manufacturer accused the landowners of rigging the price of wheat at the expense of the increasing population of urban poor. This contention naturally appealed strongly to empty stomachs, and therefore was widely supported by the very class which the protagonists were exploiting. The argument that the price of wheat might, in fact, represent a just price, and that the remedy lay in applying the same principle to the wages of the industrial worker was not considered. The political struggle culminated in a victory for the industrialists led by Cobden and Bright, and the Corn Laws which had fixed the price of wheat and controlled imports, were repealed.

The result of this resounding victory for the industrialists with their policy of free trade and *laissez-faire*, was far-reaching. From its position as the basic industry of England and the primary concern and occupation of Englishmen which must be safeguarded at all costs, agriculture now became the servant of industry. Whereas in the past the town existed to serve the needs of the surrounding rural community, the country was now subservient to the town and was to an increasing extent controlled by urban interests. Husbandry came to be regarded, not as a way of life, but as a potential source of profit, or, in other words, as an industry in the urban and modern sense of the word. For this reason its practice was re-organized on the factory system of specialized quantity production, a change which logically accompanied the re-grouping of the land into larger economic units. This was affected by the eighteenth and nineteenth century enclosures which completed the process the Tudor aristocracy had begun by finally divorcing the countryman from the soil. Henceforth he was faced with the alternatives of landless labour in another's fields or of emigration to the towns or overseas.

The English village as we see it to-day in the last stages of its decay thus represents the monument of a society robbed of its birthright. As we have seen, it is no "picturesque" survival of a "primitive" past, but the ruin of an ordered and delicately balanced system which, within its limits, achieved what we have so far failed to do, an effective marriage between individuality and communism. The village with its common fields constituted in effect a single farm, for although each individual possessed his stake in the land under the "open-field system" he held it upon trust of his good husbandry and was subject to the direction of the village council or "court-leet" as to his methods of cropping. Yet, because he himself elected this council, its authority by no means constituted an arbitrary control from without. The proof of the stability and balance of this form of social order lies in its tenacious survival through many adversities from the Saxon down to the Victorian age.

Contemporary advocates of enclosure have given us a desolate picture of the state of the land under the open-field system. Because the Commoners had been battling against a hostile social and economic tide for a century or more, this is scarcely surprising, although in fairness we should allow for the anxiety of the enclosing landlords to justify their rapacity. Again, we must not forget that the Commoner and the new agronomist looked upon the land through different eyes. The former looked upon it as the source of life; consequently the emphasis was primarily upon the land itself, upon conservation. The latter regarded the land as a source of wealth, and he was therefore concerned, just as much as the new lords of coal and iron, with what could be got out of it, upon exploitation.

The Commoner's sheep, alternately folded on his arable or grazing with the communal flock upon the common, had contributed, with the rest of his livestock, to the fertility of both. This system of communal pasturage was, and for that matter still is, the only one which enables the small freeholder to support a sufficient head of livestock to keep his land in good heart. True, the system was capable of infinite development in the light of new knowledge, but the principles upon which it had been built were as unalterable as the structure of the natural world. Unfortunately, the wisdom of those principles was only apparent in the fact that after a thousand years of their application the fertility of the land had not diminished; it was not apparent in the figures of a profit and loss account, and consequently the new commercial age regarded them merely as an archaic survival and swept them away.

The dispossessed Commons of England were reduced to the abject poverty of a starvation wage, and their revolts were crushed with a ferocity which the modernist, in order to justify the illusion of automatic progress, prefers to attribute to the so-called "dark ages". All over the country new prisons and "houses of industry" as the workhouses were euphemistically called, sprang up in order to convince the "labouring poor" that their appointed lot was to do their duty in that state of life to which it had pleased their betters to reduce them. In this process of education the rulers of England were ably abetted by a sycophantic church.

It has been argued that the demands of a growing urban population were the mainspring of the enclosure movement, but this theory is hard to reconcile with the fact that, with the exception of the eastern counties, the enclosures were followed by a decline of arable acreage. Bread—"the staff of life"—was what the new poor were demanding, but the emphasis of post-enclosure farming was upon livestock breeding, food that the poor could seldom or never afford.

Despite fluctuating prices and the slump which followed the Napoleonic wars, the new farming prospered, the repeal of the Corn Laws had little immediate effect, and it was not until 1870 that, gradually at first, the long decline of English agriculture began. Only a few far-sighted men, notably William Cobbett—"The Horseman of Apocalypse" as Chesterton so aptly called him—saw in this prosperity the false bloom upon an inward rottenness. As Cobbett rode through the Shires and saw the country's wealth draining away to the new towns, leaving both the land and its people impoverished, he cursed the new economics which were so soon destined to bankrupt the home producer by flooding the market with "cheap" food imported as interest on its foreign loans; food dearly bought at the price of colonial soil fertility; food which brought riches to the transport interests, the speculators and middle-men while it ruined the overseas grower.

These financial and commercial interests were now securely entrenched at the head of affairs where they displayed in their conduct a curious inconsistency which we can scarcely fail to regard as hypocritical. Yet of this they were innocent, for hypocrisy is a conscious vice, whereas their behaviour was the unconscious product of the uneasy marriage between the fragments of the religious values of the past and the new commercial ethics whereby public life was thought to be controlled by immutable economic laws. Their private lives were governed by a rigid code of morality with which was blended a calvinistic and joyless piety, a gloomy shadow of the religion of the Age of Faith. Personal morality did not extend to a public life in which the doctrine of freedom first expounded by Machiavelli was faithfully followed, and was reinterpreted in the philosophy of the Manchester School. This specious ideology of *laissez-faire* fostered belief in the inevitability of progress and the ultimate perfectibility of man provided he was permitted complete economic freedom. Give man that freedom and progress and prosperity must follow as surely as day follows night. The hideous slums in the growing industrial areas were regarded, if they were regarded at all, as a disagreeable but transient phase in this development, so that there was no general awakening of social conscience. When the industrialist was confronted by conditions which eyes and nose refused to ignore, his conscience was comforted by the convenient theory of self-help which argued that thanks to his conception of freedom there existed no legal obstacle to prevent the "industrious artisan" from climbing to a niche in the social heirarchy. Any attempt to improve the condition of the workers by statutory regulation was stubbornly contested on the ground that it would impose an unwarrantable restriction upon the freedom of the employer to purchase his labour, like every other commodity, in the lowest market, and to use it as he pleased.

The trading policy of the nineteenth century industrialist was consistent with his social and political theories, and was well nigh incredible in its short-sightedness. As we have already seen, an expanding market was the prime necessity of the new order and the improvement of methods of transport was effected primarily to serve this end. When the advantages derived from canals and railways had been exploited to the full the manufacturer began to look further afield for potential markets, and the scramble for overseas empire began. In this, thanks to the fact that she was the pioneer of industrialism, England scored heavily, and the English manufacturer cherished the rosy dream of importing cheap food and raw material and exporting a costly finished article. For a time his conception of England as the workshop of the world, which bears so close a resemblance to the Nazi dream of a *Herrenvolk,* came true, but he overlooked the fact that he could possess no permanent monopoly of scientific knowledge, and that what he had accomplished the "lesser breeds" might also do as well or even better. The twentieth century was to prove this truth and learn its lesson in the bitter and bloody school of world war.

Two significant events remain to be noted in the history of the period; firstly, the publication of Charles Darwin's "The Origin of Species" in 1859, and secondly, the passing of the Companies Act in 1862.

We have already remarked how the influence of the Christian philosophy waned for lack of re-affirmation in the light of new knowledge, and how, in consequence religion retired from public life. Religion, like art, depends upon symbolism to convey the profundity of its meaning, and, as a result, in the nineteenth century a stage had been reached when the implications of this symbolism was almost wholly lost so that only the literal and superficial meaning was left. It became inevitable therefore that sooner or later science would discredit the literal aspect of the biblical story of the creation and the fall. This was what Charles Darwin did. Actually there was nothing in his theory of evolution (a theory not conclusively proven) to conflict with the principles of Christianity, and had it been received by a society based on the spiritual and realistic values of a living and flexible faith, the theory of his lowly origin would have inspired man with a new humility and reverence. Instead, the new knowledge came to a people who had come to regard the natural world as the product of the eternal ferment of blind, predatory and potentially hostile forces, and the story of man, himself a predatory creature, as a constant struggle to master those forces. The result was tragic. Repudiated by the church and the older generation, the

Darwinian theory was accepted the more readily by the "progressives" who could not fail to criticize as illogical the refusal to accept the facts of archæological and geological research, and consequently to believe that the Christian philosophy had been discredited by science. Lacking any spiritual basis to which it could be related, the disciples of Darwinism therefore arrogantly proclaimed their new knowledge in justification of the gospel of materialism. The spiritual principles of the Age of Faith thus finally lost their power, and the acceptance of Darwinism represents the final stage in that process already noted whereby the doctrine originally expressed by Machiavelli filtered downward through the strata of society.

It has also been observed that the pursuit of power by the acquisition of wealth, which was the practical manifestation of the materialist doctrine, was accompanied by a dissolution of that responsibility which the possession of power should involve. In the first phase of the Industrial Revolution the manufacturer had no course but to accept full moral and financial responsibility for his actions in the conduct of his business, and no matter how strained by abuse it might be, the link of mutual dependency between employer and employed could not be severed. As the revolution progressed, however, the same change was effected in industry as had already been accomplished in government, responsibility being transferred from the individual to the group. In the interests of security, the factory system tended to imitate the great primary monopoly of banking and credit which had been the diabolic *deus ex machinâ* of the New World, by coalescing into larger and larger units. The doctrine of individualism, in fact, produced its opposite. As a result of this process the individual was relieved of moral responsibility and the tie between master and man was broken. Financial responsibility alone remained until the Company Act made possible the limited liability company. This represents no less than a fictitious and inhuman entity for whose policy no single individual can be held responsible, but which enjoys all the rights of individual freedom and is now mis-called "private enterprise". By becoming a shareholder in such an undertaking, or in other words by lending his money and deriving an income from the interest on that loan, the individual was not merely free from any responsibility for the policy of the undertaking, but was also relieved to a great extent from liability for the debts incurred in the pursuance of that policy. Irresponsible power in the shape of functionless property thus became a practical reality.

Despite bitter opposition the same defensive reason that caused employers to group together brought about a similar coalescence of employed, and the trade unions were the result. By this means the bargaining power of the workers in the sale of their labour was improved. That the aim of these unions was to secure their fair share of financial gain and not to change fundamentally the principle of work and the status of the worker is indicative of the fact that the employed had now accepted the materialist motives of the employer. The Marxian forecast that the workers of the world would eventually unite and take over the control of the industrial machine from monopoly capitalism thus merely postulated a change of command and possibly a more equable distribution of wealth. It envisaged no abandonment of materialist aims or of basic change in the structure of the complex machine of production which the Industrial Revolution had built up.

By the dawn of the twentieth century the English manufacturer had discovered that his pipe-dream of an unchallenged world market was dissolving before his eyes. The industrial expansion of Europe and the New World upon the English model resulted in the creation of a colossal productive machine which demanded an expanding market and was in no way related to human needs. Masquerading as apostles of civilization and progress, industrialists scoured the so-called backward countries to find a market for their wares. The problem of providing these countries with the necessary purchasing power was solved by usury in the shape of foreign loans. In return, the debtor nation was forced to supply its creditors with food and raw material by way of interest. In order to maintain these payments the debtor was compelled to exploit his capital resources of soil fertility and natural wealth for the sake of a production in no way related to needs or which could command a just price for his labour. Thus his crops often fetched less than their cost of production even after they had been shipped half way round the globe, while frequently they must needs be burnt or ploughed into the ground in order to remind him that his hopes of prosperity lay in scarcity rather than abundance.

This process was known as "development", and the countries subjected to it became so many cockpits of rival economic powers. All over the world ancient civilizations slowly evolved through centuries of adaptation to local environment were being exterminated, while the western powers wrangled like so many vultures over their dismembered carcases. In such circumstances a great physical conflict became inevitable, and the immediate events which led to the outbreak of the first world war in 1914 had the relative significance of a match to a powder barrel. For the same reason the particular grouping of the protagonists in the struggle and the "just causes" for which they claimed to be fighting are destined to appear equally insignificant to the dispassionate eyes of future historians.

For four years mankind endured the agony of war on a scale unprecedented in human history, and its great industrial machine was geared to the production of improved methods of slaughter for which the market was assured. For this reason the Armistice of 1918 found the great financial and industrial interests the more powerfully entrenched in a world too weary to consider the reforms of a change of heart and only concerned to pick up the broken threads of its pre-war life. Thus the old economic struggle was resumed with greater bitterness and difficulty, for during European industry's absorption in war production, many of its overseas markets had developed their own manufacturing resources on the European model. The war had accomplished nothing, and the years between 1918 and 1939 brought no peace, but were merely the prolonged truce of exhaustion preparatory to a further cataclysm.

The alternating booms and slumps of trade which were the symptom of the instability of these years gave a new impetus to that process of coalescence by which, as we have already noted, the industrialist sought to protect his interests. In consequence industry took upon itself the form with which we are familiar to-day, a vast monopoly combine employing highly specialized mechanical methods of quantity production. In this respect the evolution of capitalist industry has precisely followed the course foretold by Marx, and is therefore, to pursue his argument, ripe for transference from private to public ownership. To-day this policy has many advocates and no longer appears so revolutionary for the reason, which no Socialist cares to admit, that we have already progressed half way to Socialism without achieving any marked improvement in our common lot or any advance towards a stable social order. There is in fact little to distinguish an undertaking run by a state corporation and another controlled by private interests. Both are equally impersonal leviathans and exercise a similarly inhuman control over the lives of the individuals of whom they are composed. The shareholder or executive of the private company is indistinguishable from the government official, thanks to the dissolution of individual responsibility, and cannot stand accused of tyranny or injustice. In fact the capitalist of Communist diatribe is a figure as archaic as the villainous mill-owner of Victorian melodrama ; he has disappeared leaving behind him the machine he created together with his materialist ideals. The machine continues to function because these ideals have gained universal acceptance, and for this reason also there is no clear line of demarcation between the forces of capital and labour as Marx prophesied. The higher paid employee has become a petty capitalist, and instead of a united bloc the workers are split up into a new set of class distinctions based on income level and the character, but not necessarily the skill, of their work.

Under these circumstances the transfer of the control of the means of production from privately vested interests to a State bureaucracy claiming to represent the people can accomplish little. At its best it can only secure a more equable distribution of the profits of production and a better relation between production and consumption within its sphere of government. Assuming these ends could be accomplished they would prove merely a national palliative while an economic struggle between rival State monopolies would wage more fiercely than before and with even more disastrous results. If, then, we accept the present system upon which our civilization is based, we are inevitably led to the conclusion that world control of the means of production, and this implies rigid control over the life and work of man, is the only means by which we can avert recurrent disasters and possible extermination. It becomes pertinent therefore to examine this system at its present stage of evolution, and to ponder the possibility of there being a more desirable alternative before pinning our faith in a Wellsian Scientific Utopia.

If this new world be achieved, will it be peopled by free and enlightened individuals or by slaves and ant-men? A study of twentieth century "economic man", of his work and his leisure and the trend of his thought, may help us to answer this vital question.

PART II

Now days are dragon-ridden, the nightmare
Rides upon sleep: a drunken soldiery
Can leave the mother, murdered at her door,
To crawl in her own blood, and go scot-free;
The night can sweat with terror as before
We pieced our thoughts into philosophy,
And planned to bring the world under a rule,
Who are but weasels fighting in a hole.

—W. B. Yeats.

PART TWO

Chapter IV

THE CITY AT WORK

By the year 1939 the great monopoly industries had swollen to such an extent that the number of employees in a single group was equivalent to the population of a large town and probably in excess of that of a City State in Greek civilization. They have thus achieved the stature of a small nation within a nation, and in the degree of control which they exercise over the lives of their operatives they may fairly be said to represent a microcosm of the planned State already in being. We should bear this resemblance in mind when we consider them. This control is by no means confined to the work itself for, through its Welfare Department, the industrial organization concerns itself intimately with the workers' health, feeding, recreation and home life. To this end it has established clinics, hospitals, canteens and rest homes, sports clubs, housing schemes, pensions, and has even experimented with profit sharing. The "dark satanic mills" of the last century have been superseded by airy, well-lit workshops adequately warmed and ventilated, where the worker is protected from accident by innumerable safety precautions. His regulated hours of work are shorter, his wages higher, while he is insured against loss of wages through ill-health, with the promise of more comprehensive insurance schemes to come.

These conditions represent the modern development of a new form of social order which has been called the "benevolent business", and of which Robert Owen with his New Lanark experiment of 1800-1828 was the lonely pioneer. To those who have no practical experience of work in the modern factory they appear to be convincing evidence of the development of social consciousness and a promising augury of a new age. This is precisely the impression which monopoly industry desires to create; like the nineteenth century industrialist it recognizes the fact that the wolf of material ambition is more secure when disguised in the sheep's clothing of benevolence, and it is therefore advisable before jumping to hasty conclusions to delve more deeply into the principles on which such a community is founded.

The layman's impression of a modern machine shop is usually based on a brief conducted tour. He recalls his walk down aisle after aisle of machines in buildings the size of a cathedral, the torrent of sound and the acrid reek of hot oil. He remembers nodding and trying to look intelligent as his guide attempted to explain to him the function of the machines, the capstan and turret lathes, the gear cutters and grinders, the multi-spindle drillers and borers, the hobs and automatics, all intent on the procreation of yet more delicate and specialized machines. All he could gather was a general impression of boundless ingenuity expressed in terms of innumerable wheels and shafts gleaming with oil, and spinning or recoiling tirelessly in a complex repetitive rhythm. Bewildered and overawed he is prepared to be convinced that machinery has emancipated mankind, and is ready to build Utopia on earth so soon as we have solved our social and economic problems.

It is a measure of the extent to which we are be-devilled by machines that we identify mere complexity with true progress. Almost invariably, the layman will stand rapt before the most complicated machine in the shop, watching the mechanical dexterity of its operator with awed admiration, while he will walk past the tool room, almost the last stronghold of practical skill in the factory, without a second glance.

Because the craftsman's skill resides primarily in hand and eye, his tools are merely extensions of, not substitutes for, those faculties, and therefore they are invariably simple. It is almost an axiom, in fact, that the higher the quality of the job, the simpler and more adaptable the tools. Consequently, to modern eyes, the craftsman's work is never spectacular, and is looked upon as something archaic and "primitive". Thus we measure civilization in terms of machines rather than men, and will presumably continue so to do until we have lost all qualitative standards, unless we awaken to the fact that what we are actually building is not a new civilization but a new barbarism.

The technician on the other hand obtains a more realistic view of the machine. He is aware that it is primarily conceived for the purpose of increasing the rate of output of the article to be produced and of decreasing the cost of labour and materials, a process of development which is the logical result of scientific method employed for a material end. The question as to whether the new machine will benefit the worker in particular or humanity in general is thus a secondary consideration. The worker must accept it, drudgery or otherwise, while a Publicity Department exists for the sole purpose of persuading the public by means of psychological propaganda to accept the fruit of the new process. Utopian and humanitarian claims for the machine and its product are born of this propaganda which is framed to represent material ends as means to an exalted end, not clearly defined, called "progress".

The policy of increasing output and decreasing cost operates on the principle that individual skill is costly, slow and fallible, and should therefore be replaced wherever possible by machines which are cheap, rapid, and so infallible that their rate of output can be accurately determined and thus co-ordinated within the general scheme of production planning. This elimination of human skill results in the simplification of the task of the individual worker by division of labour and responsibility accompanied by a proportional increase in the complexity of the plant. To "tool" and "set up" machinery such as this, however, involves an expenditure of considerable labour and skill before it is capable of performing a given task, so that its use is only justifiable providing the article to be produced can be standardized and turned out in large quantities. The quality and character of the product is thus ungoverned by public need; instead, public need must be conditioned to accept an article of the utmost uniformity and of little or no intrinsic value the quality of which has been arbitrarily determined by the technician. Similarly the worker, robbed of the exercise of his skill and reduced to the performance of a specialized and repetitive mechanical task, has no more control over production than the public. The small component his machine produces is probably destined to find its appointed place on an assembly conveyor in another factory a hundred miles away, so that hand and eye can find no pride in work the function of which they cannot comprehend. He is paid on a "piece-work" basis which means that a time limit for his operation, fixed by stop watch, makes the clock a remorseless and unwearying competitor. Man and the machine which is said to have emancipated him from slavery become a single mechanical unit, and human hands, serving its needs, are impelled by a subconscious automatism on their swift unvarying course. They are in fact regarded as a single unit of production, and, until a more perfect machine can be contrived which will dispense with the human element altogether, each is dependent on the other. The working efficiency of both must therefore be maintained if an even flow of production is to be assured. This is the true function of the Welfare Department. For the machines maintenance engineers and oil, for the workers medical attention, bread and circuses.

Consciously or unconsciously, the worker is aware of this impersonal and mechanistic motive which underlies all industrial benevolence; that it is concerned with his welfare merely as a productive unit, and that a change of policy or economic expediency would speedily cause his employers to consign him to the human scrap heap of the dole without more ado. With this eventuality in mind, and actuated by the same materialistic motives which pervade the whole organization, he sets out to get what he can while he may. In consequence, a ceaseless and relentless economic warfare wages between employers and employed in which both sides tend to become increasingly unscrupulous and predatory.

For example, a large industrial organization opens a new canteen for the workers, and within a few weeks 80 per cent. of the cutlery has been pilfered. The modernist would doubtless see in this a classic instance of man's predatory nature. What he would not perceive is that such predatory behaviour has actually been acquired as the direct result of the loss of individual responsibility and of the rights which accompany it. In the small family business such depredations would seldom occur, but by severing the links of mutual responsibility and respect which once bound master and man, modern mechanized industry has brought Nemesis upon its own head. Industrial discipline becomes a battle of wits, and the more the executives frame rules and regulations the more adept the workers become in the art of circumventing them. The larger the organization the greater the irresponsibility until the ultimate of State control is reached. The matter can be summed up simply. The yeoman farmer of the past knew the commandment that a man shall not gather the gleanings of his field and opened his fields to the "leasing". The modern employer, who is merely the agent of irresponsible shareholders, thinks solely in terms of profit and loss, is ignorant of the profound implications of this law, and could not, even if he would, apply it. Consequently his workers claim by unscrupulous means that which once was theirs by right.

The more actively the mass-production policy of eliminating individual skill and responsibility is pursued, the smaller becomes the nucleus of skilled workers in proportion to the total number employed. This nucleus is represented on the executive side by departments of research, design and production planning, and on the works side by foreman, charge hands, tool-makers, tool-setters and maintenance fitters. In the highly organized plant this " key personnel " bears an ominous resemblance to the queen cell of a beehive. The control of the whole organization is ultimately vested in this technocracy which regards the workers with impersonal eyes as so many units of production to be plotted on a chart. Conversely it requires a deliberate mental effort on the part of the worker before he can comprehend that the higher executive who thus remotely and dispassionately controls his destiny is in fact a creature of common human frailty like himself.

It must be acknowledged that absolute human equality is both improbable and undesirable. For if individual ability is given freedom of opportunity some will inevitably excel in the exercise of that ability. Master and man must therefore exist in any form of social order, and it is upon the harmony of the relationship between them that the stability of the system depends. Such harmony can only be represented by a mutual dependency of trust, responsibility and respect, and for these ties we look in vain in the relationship between the worker and the controlling technocracy of the present industrial system. The duties and responsibilities of both have been so sub-divided that neither master nor man exists as an individual. We are thus faced with the spectacle of an experiment in communal living the direct opposite of that of the Cistercians. The Cistercian lay brother found his freedom in individual responsibility for the perfection of his work to the benefit of the community; the modern worker has lost his freedom, together with his right of self expression in the pursuit of individual gain. The modern industrial organization, in fact, is held together by no common purpose whatever, but by the individual striving of ant-men, each pursuing his private goal of material ambition, engaged, as William Morris expressed it, " In the reckless waste of life in pursuit of the means of life." We should hesitate before we submit to the clamour of the scientific progressivists and permit the extension of such a system on a national scale.

The freedom of ability which we have forfeited is the essential condition of craftsmanship, a term that has been much abused. A specious "Arts and Crafts" movement has made the craftsman a museum piece, and has derived great profit by fostering a luxury trade in second-rate goods on the score that they are "hand-made". The spectacle of a middle-aged spinster in the carefully restored cottage of a "show-village" producing shoddy loose-woven cloth on a rickety hand-loom such as no craftsman would use affords a good example of this form of travesty. A common misconception is thus created which no doubt partly accounts for the fact that the progressivist habitually dismisses any champion of craftsmanship as a romantic traditionalist, a shallow philosopher who, unable to face the problems of the present, resorts to a nostalgic evocation of a past which never existed in fact.

Craftsmanship, however, is not a synonym for Artfulness and Craftiness, neither is it the exclusive attribute of a Chippendale or a Sheraton. It is a fundamental principle, a qualitative conception of work as an integral part of the art of living, and it is capable of informing all work from making a motor car, a clock, a chair or a plum pie to digging a garden or building a rick. It is the seed from which all art has flowered and its products are the memorials by which civilizations are ultimately judged. Its exercise does not involve the abolition of science and machines, but it indicates their proper function as an aid to, and not a substitute for the skill of hand and eye.

It is this skill which we have seen fit to banish from our world by denying the worker the freedom of his ability, the freedom to express his creative instinct which is his supreme natural gift. Even though we may gain the whole world in pursuit of our material aims, if we continue to deny to the common man this right, then we shall lose our souls.

Because the true craftsman is unambitious in our modern sense of the word he has had little or no influence on the march of events, and, because he is the exemplar of a dying tradition, that influence grows yearly less. Only his work speaks for him, for although he is deeply conscious of a world quite out of tune with the principles on which his life has been based, he is seldom articulate in the literary sense. Fortunately there have been exceptions, not least among them being George Bourne, whose

"Wheelwright's Shop" has become a minor classic. His own words at the close of this book make a more fitting conclusion to this chapter than any words of mine, since they cannot be lightly dismissed as unrealistic romanticism. Forced by economic necessity to introduce mechanized methods into his shop he writes:

"There in my old-fashioned shop the new machinery had almost forced its way in—the thin end of the wedge of scientific engineering. And from the first day the machines began running, the use of axes and adzes disappeared from the well-known place, the saws and saw-pit became obsolete. We forgot what chips were like. There, in that one little spot, the ancient provincial life of England was put into a back seat. It made a difference to me personally, little as I dreamt of such a thing. 'The men', though still my friends, as I fancied, became machine 'hands'. Unintentionally, I had made them servants waiting upon gas combustion. No longer was the power of horses the only force they had to consider. Rather, they were under the power of molecular forces. But to this day the few survivors of them do not know it. They think 'unrest' most wicked.

"Yet it must be owned that the older conditions of 'rest' have in fact all but dropped out of modern industry. Of course wages are higher—many a workman to-day receives a larger income than I was ever able to get as 'profit' when I was an employer. But no higher wage, no income, will buy for men that satisfaction which of old—until machinery made drudges of them—streamed into their muscles all day long from the close contact with iron, timber, clay, wind and wave, horse-strength. It tingled up in the niceties of touch, sight, scent. The very ears unawares received it, as when the plane went singing over the wood, or the exact chisel went tapping in (under the mallet) to the hard ash with gentle sound. But these intimacies are over. Although they have so much more leisure men can now taste little solace in life, of the sort that skilled hand-work used to yield to them. . . . In what was once the wheelwright's shop, where Englishmen grew friendly with the grain of timber and with sharp tool, nowadays untrained youths wait upon machines, hardly knowing oak from ash or caring for the qualities of either. And this is but one tiny item in the immensity of changes which have overtaken labour throughout the civilized world. . . . That civilization may flourish a less civilized working-class must work."

CHAPTER V

THE CITY AT HOME

WHEN at evening the factory sirens blow and workshops and offices disgorge their jostling crowds, the worker's concern is to leave his work as fast and as far as he is able, and, by 'bus, car or cycle he journeys toward the suburbs on the city's perimeter. This understandable desire to escape from the conditions created by industrialism, and its repeated defeat by its own result is the factor responsible for the city's haphazard growth.

At night its blackened late Georgian or early Victorian core is left to the poorest classes who inhabit the mean streets and courts of back-to-backs which still surround the older factories and mills. As we walk outwards from this centre we next pass melancholy rows of detached villas which date from the same period and once housed the first mill owners who deemed twenty minutes drive in a carriage sufficient escape. When they were built their windows looked out on green fields, but now the city has engulfed and wealth abandoned them; stucco peels, bricks crumble and railings rust; dark overgrown shrubberies drip on the weedy gravel and darken the lower rooms; they have become tenements or cheap boarding houses. Beyond them the road leads to the first suburb where an ancient church and a few cottages survive incongruously to reveal that it was once a village outside the city boundary. In the first years of this century it became a smart retreat for the more affluent of the city fathers whose residence lent to its name a social cachet in the city which would have amazed its earlier inhabitants. Here, protected by a green belt from the ugly source of their income and from contamination with the lower orders, wealth dwelt for a time secure. But alas! the motor 'bus soon violated their sanctuary, and they were forced to abandon their Edwardian villas in the half-timbered style to the siege of the aspiring tradesmen's semi-detached homes which advanced upon them inexorably rank upon rank. Finally this swollen suburb with its improved communications, its gas, electricity and water mains attracted the enterprising industrialist the coming of whose factories inaugurated a fresh cycle of expansion by flight.

This brings us to the great belt of housing estates and ribbon development, a hinterland that is neither town nor country, which has come into being in the past two decades and has drawn another ring of villages within the city's widening orbit, to the great profit of the land speculator and to the ruin of local agriculture. Here there is plenty of evidence to show that yet further expansion is imminent so soon as the present war-time restrictions on building are removed. Already industry has established itself, and true to the example set by their Victorian and Edwardian forebears, the captains of that industry have removed themselves from its mundane environs to desirable country residences in the neighbouring village. The narrow road to this village needs widening so that the Bentleys and S.S.s may speed home safely; then the bridge over the millstream will be rebuilt to take the weight of the Corporation double-decker in the wake of which will follow the cinema and the chain store. The mill itself will acquire a new lease of life as a bathing lido, and in a year or two the village will experience all the joys of urban "amenities" and the rates will be trebled. This, briefly, is the process of infiltration by which our new material civilization of cities has engulfed the older order represented by the village. The extent to which it has already succeeded may be judged from the fact that out of a total population of 42,000,000 in these islands, 35,000,000 live in urban areas.

Whereas the unit of the past was the village, the unit, or rather agglomeration, of the present is the housing estate, so that to draw a comparison between the two is not invidious. Where the village expressed individuality and regionalism in the diversity and character of buildings whose only common quality was the conception of craftsmanship which built them, the housing estate presents the bleak uniformity enforced by mass production. Where the village reveals a communal unity in its carefully chosen site and the intimate relation of each part to the whole, to each other, and to the work of the fields and its attendant crafts, its modern equivalent sprawls witlessly over hill and valley displaying a blank and alien individualism incapable of concession to its surroundings and lacking any common purpose. Our materialism has in fact produced a social organism which is the precise opposite of that which it has replaced; for a community of free individuals it has given us a dormitory of economic thralls, for "wholeness" and self-sufficiency, "separatism" and isolation. It furnishes no evidence of that solidarity of the workers which Marx forecast; on the contrary, it reflects the new system of class distinction based on earning capacity. The foreman and the clerk live in semi-detached rows near the main road to the city, while the higher salaried technicians and executives have their detached homes in the tree-lined avenues near the golf course. Their houses, priced like their motor cars to meet different income levels and grouped on this basis, encourage the creation of class barriers as rigid as those of the last century.

Having followed the worker home and so traced the development of the city as a whole, we can now enter his front door and so study the individual unit. Compared with his predecessor in the Victorian slum, his living no less than his working conditions present superficially a striking improvement. His new house is light and roomy, there is a bathroom, a constant water supply and main drainage. Electricity and gas give him heat and light for the flick of a switch or the scrape of a match. The grid system and mass production have between them mechanized the home, and ingenious machines preserve and cook his food, clean the carpet, keep the time, wash and iron the clothes, heat the bath water, shave him and even dry his wife's hair. Her duties have still further been relieved by the science of food preservation which makes the can-opener the most useful implement in her kitchen of chromium and stainless steel. Even her task of shopping has been largely alleviated by means of the telephone and modern delivery facilities, so that bread, milk and groceries arrive, like the newspaper, with the unremarked regularity of sunrise.[1] She has thus been "emancipated" by science from the exercise of a craft which shares with that of agriculture the distinction of being the oldest in the world. The story of this emancipation carries us back once more to the home of a pre-industrial past before the majority had become infected with the virus of materialism, and it enables us to make a more fundamental valuation of the modern labour-saving home than is possible by comparing it to the Victorian slum dwelling which was but an early manifestation of the same system.

The kitchens of Sulgrave Manor in Northamptonshire, or of the Fleece Inn at Bretforton, are accessible examples of the

[1] It should be emphasised that this is a review of the conditions obtaining in 1939!

domestic hearth from which, in a comparatively brief space of time, we have travelled so far. Of these, one is already dead, but, for those of us who have sufficient imagination and humility to learn from the past instead of merely sentimentalizing over it, they are not merely archaic museum pieces, " picturesque " relics of the primitive to be goggled at by urban sightseers. The furniture which the country carpenter fashioned two or three hundred years ago is as sound as the day it was made, and the mellow bloom which table and dresser have acquired from the use and care of generations reflects the gleam of copper and pewter. The blacksmith, too, has left about this hearth in skilfully forged crane and ratchet hook, trivet, fireiron and shining spit, a finer memorial than any tombstone. The bread-peel leans beside the door of the bread-oven, like a symbolic staff of life; the hooter waits to mull the ale when nights are cold, and wool-comb, winder and spinning wheel await the hands that once carded and spun. All the emblems of work in field and garden find their place about this hearth, shepherd's crook and quilted smock, flail and harvest bottle, skimmer, apple-scoop and spice-box; bacon rack and seasoning herbs, marjoram, thyme, chervil, winter savoury, balm, comfrey and rosemary. All these things and many more, in which use and beauty combine to produce intrinsic worth, speak and speak eloquently of a vanished household whose conception of the home was fundamentally opposed to our own. The keynote of this conception, like that of the village of which the home was a part, was self-sufficiency, a unity of life and work in indivisible association wherein the kitchen, as the workshop of woman's craft and scene of man's rest after labour, was the centre and focal point of manifold activity.

The growth of materialism, which, as we have seen, filtered downward through a strata of society brought with it a drastic change in the status of the home. Instead of a workshop wherein the art of living was practised it became a vehicle for the display of wealth, a measure, in fact, of man's success in the pursuit of power. In the nineteenth century this baneful influence reached cottage and farmhouse where it was made manifest in the " front-room ". The hearth and all that it represented were sacrificed to this gilded calf of tawdry splendour, to faded plush, veneer furniture and useless bric-à-brac, to cheap and worthless imitations of the rich man's finery. Fallen thus from its high estate the kitchen became, like closet and bathroom, one of the " usual offices" discreetly hidden from the eyes of callers, while in accordance with the new conception of gentility its function became the sole responsibility of servants wherever possible. Deprived of her workshop the Victorian housewife of the middle classes adopted the rôle of an odalisque, her finery and unlimited leisure, apart from child-bearing, being requisite as a measure of her husband's success in the economic battleground. Sooner or later human nature was bound to react against this unnatural, cloistered existence of inactivity and boredom, and the twentieth century soon found her clamouring for suffrage and higher status in commercial life.

The last war brought about a certain levelling of income; the majority of the middle class could no longer afford to leave the kitchen to the staff behind a green baize door, while a rise in the standard of living of the poor, and extensive re-housing, enabled a greater number to emulate the standard set by the more favoured. Thus the era of domestic service came to an early end, but the old tradition of the home had been slain irrevocably, and the housewife returned to her duties reluctantly, her hands no longer informed by the pride and skill of an inherited traditional craft. Modern industry, always quick to develop a potential market, solved the problem by providing her with machines and scientific aids in place of servants. It gave her food that needs no cooking, synthetic concentrates and bull-in-the-bottle in place of the herbs, spices and seasonings whose secrets her great-grandmother knew; it left on her doorstep the steam-baked loaf of white flour from which its roller mills had extracted all the nourishment of bran and wheat germ to sell in the form of patent foods. In place of the once great company of English cheeses, Cheshire, Blue Vinney, Stilton, and Double Gloucester, to mention only a few, it brought her from the other side of the globe a so-called "Cheddar", fit only to bait mousetraps, or a curious soap-like substance wrapped in silver paper. Instead of the fresh products of home, field, and garden, it substituted frozen meat, stale vegetables, fruit and cereals wrung from remote colonies at the terrible price of soil exhaustion and sold them to her for less than they cost the farmer to produce. It gave her imported butter that resembled margarine, and thus made the substitute more palatable, it gave her pulp jam and ready-made cakes, and it filled her cellar or cupboard under the stairs with beer and wine which a countryman accustomed to the home-brewed liquor of the past would have spat out with a wry face.

To the domestic, as distinct from the culinary side, science and industry in partnership have also contributed liberally in addition to the electrical appliances already mentioned. To replace copper, brass and pewter they offer metals whose lustre cannot tarnish and whose cold surface of perpetual brilliance can therefore never reflect the bloom of use and care. For the "front room" complex, which has become a power that now dominates the whole house, they cater in full measure with cheap ornaments and pretentious suites of factory made furniture, whose ill-fitting drawers and doors, shoddy workmanship and green timber is masked by a synthetic finish and meaningless embellishment. In the process of substituting the genuine by the mass-produced counterfeit the industrialists have become masters in the art of disguising materials. The oak panelling and plaster ceiling of a dining-room turns out to be embossed paper; cheap plastics or metal pressings resemble mahogany. The stone fireplace and the marble bath are likewise spurious, tapestry hangings and covers are printed copies, while the tiled floor of hall and kitchen, like the parquet in the front room are sheets of linoleum.

Such a compendium of the victories of science in the home could be prolonged to wearisome length. The important point about them is to note that they were not won without effort. Despite all the ravages of the nineteenth century the Englishman still preserved some vestiges of that natural good taste which was a legacy of the sense of fitness that his forebears possessed. Under the banner of progressive beneficence the propaganda of mechanized industry went to war against notions they termed conservative and old-fashioned. They proclaimed themselves the heralds of a new Utopia, knights errant bent on rescuing the housewife from the dragon of drudgery, while simultaneously they subtly played upon the snobbery, ambition and fear of rivalry which the front room complex engendered. By these means they secured a market for their shoddy wares and dealt a death blow to the critical faculty of a people which was already undermined by uncreative work. Not only outwardly but inwardly, the worker's suburban home resembles fifty others on the same estate and thousands more elsewhere. Because the house and its contents are the standardized product of machines, they do not possess, neither can they acquire, any of that intrinsic value which a combination of individuality, beauty and fitness for purpose can alone bestow. The essence of this home is its separateness, instead of the focal point of the innumerable intimate associations of life and work, it has become a dormitory and a refuge for owners who could not even if they would, distil upon it one spark of that corporate spirit which once informed the cottage of the poorest peasant.

The worker only returns to these surroundings to eat and sleep, but his wife, unless she works also, has her fill of them. She seldom has a large family to occupy her since the lure of the small car, the radiogram and the refrigerator have made babies an indulgence to be indefinitely postponed. Consequently her hands, freed by friendly science, lie idle, and she becomes as bored as her Victorian middle-class forerunner. She develops a nervous complaint which a member of the medical profession has appropriately called " Suburban Neurosis ", and which is not only brought about by boredom, but by fear. This fear is induced by economic insecurity, a factor which seldom wrinkled the brow of the complacent middle-class of yesterday. In the modern suburb, however, thanks to the hire purchase system it is an ever present spectre.

This ingenious system of usury was evolved by the manufacturer in the course of his ceaseless quest for new markets in order that the consumer could be encouraged to acquire goods beyond the reach of his purchasing power. The vicarious joys of vieing with the neighbours in the acquisition of faster motor cars or bigger and better radiograms, more opulent furniture and even larger and more select houses, these were the tempting lures skilfully offered in the usual disguise of benevolence. All too often the victim swallows the bait, forgetting that usurers never fail to exact their pound of flesh. For should the worker's income fail through loss of employment, always an imminent danger in an industrial system which inevitably progresses in a recurring cycle of booms and slumps, the mask of benevolence quickly drops and the hire purchase company withdraws its wares, having exacted an exorbitant sum for their hire. If, on the other hand, he is more fortunate and maintains the payments, the worker actually fares little better, for by the time the goods become his own they are worn out or obsolete, and in effect he has been paying, and paying heavily, for the doubtful privilege of using them for the period of their useful life. The manufacturer in his own interest ensures that that life is short, for if he made heirlooms his market would become saturated and his machines cease to hum. This applies to the suburban house itself no less than to its contents. Flung up from cheap materials by a speculative builder with an eye for quick returns, its "limey" bricks and breeze blocks will crumble, its green timbers warp and plaster crack long before its occupants have paid their last instalment.

Thus the much vaunted "amenities" of the modern city are merely tinsel glitter, and the individual urban home no less than the city of which it is a unit, represents the art of living in decay, the false gods which materialism has set up, and the sterility, loneliness and spiritual impoverishment which their worship has bred. For not only is the house and its contents intrinsically worthless and graceless, but the whole crazy impermanent structure rests on the quicksand of usury. Small wonder that economic man finds in his home life little solace and no escape from the din of his machines and the clamour of the counting house, but demands the distraction of mass entertainment.

Chapter VI

THE CITY AT PLAY

In the rural civilization of the pre-enclosure past the peasant of the village community created his own amusements. These were principally associated with his festivals or holy days which were of religious origin, fundamentally pagan, but later Christianized. According to contemporary records these represented a total of eight weeks in the year in the twelfth century. The majority were based on ancient traditional ritual and ceremony intimately associated with seasonal change and the work of the fields, and tacit acknowledgment of man's dependency on nature. Perhaps the most ancient were Candlemas (February 2nd), Rood Day or Eve of May, Lammas (August 1st) and All Hallows, for such a division of the year at May and November with two cross-quarter days is neither consonant with sowing and reaping, nor with solstice and equinox, but with the commencement of the two animal breeding seasons. It is therefore probable that they originated among a purely pastoral people. To these were added the solar festivals of Christmas and Midsummer, the agricultural ceremonies such as those associated with haysel and harvest, and numerous others of regional significance. Of all these some survive as public holidays, others merely as names on the calendar, but only the ceremonies associated with Christmas survive on a national scale, and even these have to a great extent lost their meaning. These seasonal festivals were not only inseparably interwoven with man's life and work, but they in no way resembled the joyless sanctimony of the Sabbath which Puritanism imposed. The terms holy day and holiday were then synonymous, and it was not until materialism clouded man's vision that the significance of the festivals was lost and joy could no longer be reconciled with faith, pleasure with piety. On the contrary they were the occasions of games, dances, feasts, plays and songs, echoes of whose richness linger in the memory of old men and in a few tattered but tenacious survivals of flesh without spirit such as the Helston Furry Dance (to quote the best known example), the Mummers Play, versions of which still exist in a few villages, the Horned Dance of Abbots Bromley, the Besant of Shaftesbury, or the Hocktide ceremonies at Hungerford.

For many years following the Renaissance many of these annual feast days, together with those of the trade guilds, continued to be observed in the growing commercial towns. But the traditions which had inspired them became gradually obscured, and their significance lost, because their validity depended upon their intimate association with conditions of life and work to which the new material civilization of cities was fundamentally alien. To the new urban poor they were no longer holy days which celebrated and sanctified the fruitful partnership between man and nature, they were holidays which, like the cheap spirits they so freely imbibed, afforded temporary release from drudgery and squalor. The Nirvana they demanded called for stronger meat than strolling player or country dance could provide, and in the train of their drinking bouts came such brutal sports as kickshins, back-swords, bull-baiting, and cock-fighting, followed by a rabble of card sharpers, gamblers and bullies. These degenerate forms of amusement gave the growing Puritanical element legitimate grounds for repressive action, and one by one the great fairs, feasts, mops and wakes disappeared or survived as a mere shadow of former substance. The history of the famous Cotswold Games on Dovers Hill, near Campden, which Shakespeare and Ben Jonson knew, affords a typical instance of this process of degeneration. First suppressed by the Puritans of the Commonwealth, but revived by the redoubtable Captain Dover at the Restoration, they continued to be held until 1852 when, thanks to an unsavoury reputation lent by the invasion of gangs of rowdies and thugs from the Black Country, they came to an unlamented end.

Meanwhile, in the everyday life of the village community, the introduction of the organ into the church severed the last remaining link between the sacred and the secular. The village band with fiddle, viol and serpent, equally at home in church gallery or inn taproom, was henceforth banished from the church. Most of the early organs were of barrel type, and the Worcestershire village history previously quoted relates how, in 1834, William Southwell, who had previously been in charge of the band, was paid £1 for "turning the organ".[1]

[1] According to the parish records a yearly payment of 5/- had previously been made to Southwell "for strings for the viol and bass viol". Subsequently the accounts record payments of £16 17s. od. for repairs to the organ over a period of three years.

Thus even in the church the machine was replacing the man, and it is easy to imagine this old musician's wry face as he assayed his new monotonous task. Soon his despised instruments were to be banished from the inn also. "This", wrote Alfred Williams in the introduction to his collection of Upper Thames folk songs, "was the most unkind and fatal repulse of all. It was chiefly brought about, I am told, not by any desire of the landlord, but by the harsh and strict supervision of the police. They practically forbade singing. The houses at which it was held, i.e. those at which the poor labourers commonly gathered, were marked as disorderly places; the police looked upon song-singing as a species of rowdyism. Their frequent complaints and threats to the landlords filled them with misgivings; the result was that they were forced, as a means of self-protection, to request their customers not to sing on the premises, or, at any rate, not to allow themselves to be heard. The crestfallen and disappointed labourers accordingly held their peace. The songs, since they could no longer be sung in public, were relegated to oblivion; hundreds have completely died out and will be heard no more.

Thus the countryman's fiddle, relegated to loft or outhouse, is something more than a sad piece of detritus, a tangle of splintered wood and broken gut; it is the symbol of a spring of joy ruthlessly stopped up. Dispossessed by the enclosures, transported, herded in slums, half-starved, ground down by a barbarism such as he had never known, the peasant still sang until "progress" clapped its hand over his mouth.

Throughout the nineteenth century, although other and less brutal sports survived as common forms of amusement, urbanized man tended more and more to rely for his diversion upon the professional efforts of others rather than upon his own, and in consequence entertainment became an industry. Though his material standards of living were beginning to show signs of improvement, his surroundings and his conditions of work were becoming increasingly monotonous and stultifying with the result that he still demanded the antidote of sensation. This was the golden age of the circus, the music hall and the provincial theatre with their wild beasts, freaks and high wire acts, their racy humour and blood curdling melodrama. Although these institutions catered for a degenerate taste and marked the beginnings of specialization and commercialism in entertainment, at least the performers constituted communities bound closely together by that sense of custom and tradition which mutual skill and pride in craft invariably creates. Yet their heyday was all too short, for, like many another craft before them, they were soon to be subordinated to the machines of scientific monopoly industry, in this case the cinematograph (later to find a voice in the selenium cell), the gramophone and the radio transmitter and receiver. To these television will soon be added. Here it is relevant to remark once more that these inventions were not developed primarily by the stimulus of public demand, but as a potential source of profit; that they benefited neither the scientists and craftsmen who perfected them, nor the community, but the financial interests which sponsored and exploited them in the guise of benefactors. As a result of this mechanization, popular entertainment became an industry which quickly progressed on its inevitable course towards centralized technocratic monopoly represented by the film company, the cinema ring and the Broadcasting Corporation.

When metropolitan man goes out of an evening he pursues distraction, he does not desire to think but to forget; to forget for a while the monotony of his work and the spiritual barrenness of his suburban home. He does not have to journey far to find his temple of escapism, for the cinema is swifter than church or school in following his flight from the city centre. Here in the hot darkness he can wander for a while in that dream world of extravagant wealth, eroticism and sensation, which is the film company's translation of material wish-fulfilment. The mass production amusement factories of Hollywood, whose "stars" have killed the strolling player, know the recipe. Take plenty of rich romance, palm trees, moonlit beaches, soft music; season with the spice of sex intrigue; add an equal amount of sensation, a spectacular car or 'plane crash, a shooting affray, or preferably something unusual to tickle the filmgoer's jaded palate such as an earthquake or a stampede of elephants; mix well and decorate with the snobbery of wealth, millionaire apartments, marble swimming pools or luxury yachts, and the result will be a product that cannot fail. For a brief hour or two every man in the audience will become a Clark Gable, while the glamorous half-naked blonde whom he clasps in his arms under the swaying palms will represent the dream of every neurotic little suburban housewife or jaded typist who bleaches her hair or paints her face and nails in emulation.

This may seem to be harsh criticism which ignores the bolder and more imaginative experiments which film studios have made in recent years, particularly in the documentary or semi-documentary field. Yet no matter how great the individual ability of producer or camera man, he can never succeed in overcoming the defects inherent in a mechanical medium controlled by a monopolistic organization operating on a profit basis. Emphasis will be laid in a later chapter upon the unreality of much that we assume to be real in the modern world due to its foundation on relative and therefore abstract values. Much of the art of the cinema consists of creating, by means of models and trick photography, an illusion of this illusory reality so that the film, so far from being the "slice of life" which it purports to be, is really an abstraction based on an abstraction, and is thus further removed from reality than the wildest fantasy. It is therefore at the opposite pole to the true principles of art. For art does not, or should not, hold a distorting mirror to the falsity of contemporary life, it seeks to portray some aspect of a reality that is impervious to "the bank and shoal of time", and this quality of timelessness is the measure of its greatness.

Again, owing to the complexity of film technique, responsibility for the product, as in any other modern commercial organization, is divided, so that the film, no less than the mass-produced motor car, is an impersonal product bearing no stamp of individual craftsmanship. Finally, owing to the financial factor, universal popularity over the widest field is the dominant consideration when a picture is projected, so that the result usually conforms to the lowest common denominator of intelligence. Consequently, there is a continual tendency for the standard of films, instead of rising, to fall toward the abyss of infantilism.

This debased form of entertainment not only has an enormous standardizing influence on the public mind, but it is salutary to reflect that it exercises this influence on a world wide scale, and so represents, in fact, the first manifestation of an international "folk-culture". It is thus extremely significant as a mirror of the present ideals of world civilization, and as a portent of things to come. The international character of cinema "culture" is nowhere better expressed than in its music, which has become the accompaniment of the modern dance, that dreary cosmopolitan shuffle that knows neither grace nor gaiety. In the course of an evening spent at a dance it is possible to hear travesties of tunes plucked carelessly from their roots in a dozen regional civilizations of which they were once the expressive flower; a Viennese waltz, a Negro spiritual; Spanish tango; Cuban rhumba; even the haunting folk-music of the Polynesian, Tzigane, Celt and Gael does not escape the predatory net of the mass-production music makers.

Before the swift advance of this brazen mechanical voice the theatre is in full retreat and scarcely exists as a popular form of entertainment outside the metropolis, which has become its stronghold, and to which devotees of the drama must make pilgrimage. Yet even that drama, if it is to be popular, must corrupt itself on the model of the cinema's photographic "realism" by the use of sensational action, the impoverished speech of everyday life, and elaborate scenery, so that no demands are made on that intellectual collaboration of the audience which is imagination. The dramatist of the past played upon this imaginative response by means of the verbal magic of poetry and soliloquy in order to clothe his simple stage and to people it with creatures of more than mortal stature. For the common man in pursuit of distraction such a mental effort is now neither possible nor desired, and for this reason plays that once charmed his "primitive" forebears in tavern yards he considers "high-brow".

In the realm of sport urban man's desire for sensational distraction without the effort of personal contribution is equally evident, for it is estimated that Association football interests six times as many English people as any other form of sport. Of these only a small minority are active participants, the rest are content to be spectators at the great contests between highly paid teams of specialists which, in the importance attached to them and the vast crowds which they draw, bear a significant resemblance to the gladiatorial games of a decadent Roman Empire. Next in importance to football come horse-racing and the new semi-mechanical sport of greyhound racing, in which the rôle of the devotee is also passive, and where most of the sensation is provided by the gambling interest for which the totalisator machine caters so adequately. Even among the time-honoured pastimes of the public-house there is evidence of decline. The skittle alley which once resounded to the rumble of "woods" and rattle of falling pins is now empty or ruined, while even the dartboard, though still popular, is menaced by the electric pin-table.

The inn itself, which once shared a place with the church at the centre of village life as man's parliament, playground and solace after labour, has been transformed. Floodlit and neon-signed, its suburban counterpart in brewer's Tudor preserves no shadow of the old community life. A drink shop, its many bars with their ascending scale of prices for the same liquor emphasise the new suburban system of economic class distinction. The cocktail bar where the sales manager from the "Hollies" can drink his whisky on the way home from golf is out of eye and earshot of the "private" where the tool-room foreman from Acacia Avenue consumes his bottled beer. He in his turn is decently segregated from the "public" where machine-hand and labourer down pints of washy "half-and-half". These large unfriendly rooms are little conducive to talk and laughter, and such as there is, is all too often overwhelmed by the blare of radio, a constant distraction, like a dripping tap, to which no one listens.

These are the pleasures of the leisure hours of twentieth century economic man, and together they represent that insatiable desire to be entertained, which is due to his inability to entertain himself. This inability is once more revealed in his holiday excursions further afield when, to avoid the boredom and loneliness of the country to which he returns as an alien, he brings his urban pleasures with him and feels at home only in the gregarious atmosphere of the holiday camp and pleasure resort.

From this brief summary it should be evident that the growth of scientific materialism, by substituting machines for individual ability and uniformity for diversity, has brought about as great a change in man's leisure as it has in his work, and for precisely the same reasons. This is a fact which those Utopians who look forward to an age when machines, by absolving man from labour will give him increased leisure, would do well to ponder. For their theory that the spirit of man will achieve its highest expression when it has been freed from the "slavery" of work is manifestly a fallacy since that spirit can only express itself in those creative activities which constitute that art of living of which work and leisure are not separate parts, but facets of the one whole. If that creative freedom be denied and lost, though he work but one hour a day, man will not be freeman but slave, and just as the efficiency of his machines will determine his hours of liberty, so they will be the measure of his boredom. For boredom is the disease of minds in which the creative impulse has been atrophied, and for which sensation is the only drug.

It is in this hotbed of restlessness and boredom that the seeds of war germinate.

Chapter VII

THE CITY THINKS

The last three chapters have presented a synopsis of the dominant trend of modern civilization as it affects the average man. To treat so vast a subject in so small a compass inevitably involves the broadest generalization. The "average man" whose way of life I have sketched is a saw-dust figure, for I have ignored those qualities which make each man an individual. Single out the most unpromising figure in an urban crowd, probe deeply enough beneath the hum-drum exterior, and there would be found some channel through which the positive good in his individual nature finds an outlet, even though it be inarticulate; a love of music, perhaps, or of the country; a passion for gardening, or for some other creative hobby. Once tap that source and the individual wakes to life.

The tragedy, however, is that the conditions of life which I have outlined represent a collective "norm" to which, if material progress pursues its present course, an increasing proportion of the world's population must conform. Unless they do so the planned machine state cannot function. Consequently, individual channels of expression become "pointless" indulgences detached from the business of living; luxuries only to be tolerated so long as they do not conflict with a social pattern arbitrarily prescribed. The fact that it is so prescribed is due to the fallacious assumption that the present form and scale of material progress is inevitable and immutable whereas in fact, being a product of the human mind, it is not only questionable, but sensible of change.

This implicit belief in automatic progress is a product of the materialistic philosophy which has become the accepted foundation of our thought. Materialism is essentially objective and this, to use a dictionary definition, means that it "lay stress upon that which is external to the mind, treats of outward things and events rather than of inward thought, or exhibits the actual facts not coloured by the opinions or feelings." Consequently it is these outward things which, for the materialist, constitute "reality" and "truth". Unfortunately, however, these objective standards, no less than man himself, are subject to the flux of time, and on this account their value is changeable and relative. Religious thought, on the other hand, is subjective because it postulates truth as an absolute which, because it exists in a reality outside time, is therefore changeless and eternal. Every human mind is susceptible to an inkling of this truth, although it cannot be logically defined or enunciated by the intellect, because such a process involves the use of reason which can only express itself in terms of objective and therefore relative value. Man's religion and his art represent his efforts to translate the absolute in terms of the relative. In them, objective "facts" are used as symbols which together convey more than the sum of their parts.

In the introduction to this book, knowledge was defined by analogy as a set of tools, wisdom as the ability to use them. We can now carry this definition further by saying that knowledge is concerned with objective and relative facts, while wisdom represents our ability to associate and reconcile these facts with our perception of absolute truth. The mythology of the Saint and of the Golden Age postulate the unattainable ideal, a perfect marriage of knowledge with wisdom in the individual life and in the life of a whole people. In practice, the achievement of the individual life or of a whole civilization is dependent upon the degree of harmony achieved. This harmony can never be static since men's minds are constanting pursuing knowledge no less actively than truth. Owing to this material development relative values are subject to continual fluctuation and, since it can only be expressed in terms of these values, absolute truth, though itself a constant, requires continual re-statement. If such affirmation is not made knowledge outruns wisdom, its symbolism loses its significance to become a dogma, and the balance is overthrown. So far all man's experiments in civilization have fallen owing to their failure to maintain this balance, and there is evidence that our own attempt is following a similar cycle.

The mediæval period in Europe is called the Age of Faith because it was based, not on the objective realism of reason, but upon the subjective principles of Christian religious belief which were accepted unquestionably as the absolute basis of life, the unalterable standard of right and wrong, and the supreme arbiter of both king and peasant. These principles informed every aspect of the waking life to produce that organic wholeness of living, that indivisible association between labour, leisure and home life, between the individual and the community, piety and pleasure, which characterized the self-sufficient organization of the Cistercian Monastery and the pre-enclosure village system. These experiments in communal living represent no Golden Age as some romantic mediævalists would have us believe, but they are the finest flowers of an age of ferment, violence and infinite promise when for a time the vision of a united Christendom fired the hearts of men only to vanish all too soon beyond recall. The first chapter of this book attempted to discover the reason for this eclipse, and to trace the emergence of a new and fundamental change in human outlook out of the corruption and secularization of the church on the one hand and the arrogance of the Renaissance nobility on the other. For lack of re-affirmation, custom, creed and ceremony which had been but the outward symbol of inward and spiritual truth lost all but their literal meaning and moral law could therefore no longer command minds eager in the pursuit of knowledge as power. Realizing this, the priesthood sought to enforce a dogma and restrict the growth of knowledge by embarking upon that disastrous rule of superstition and fear enforced by stake and faggot which contributed to their downfall.

In the course of this, the Christian conception of the universe as an ordered and harmonious creation and of man as an organic part of that harmony, or as its disruptor (his rôle being dependent upon his use or misuse of the gift of free-will), gave way to the Puritanical idea of predestination, or to the purely materialistic view of the universe as mechanism. In neither philosophy could man be held solely responsible for the consequences of his own actions, so that their acceptance destroyed the tragic sense along with the realistic recognition of good and evil. He became the slave of time, in the one case a puppet, and in the other an automaton. The difference is slight, and the result was the same. It produces, as we have seen, an arrogant and predatory individualism, a conception of freedom that destroys freedom. It also rejects all that cannot be translated into the terms of the mechanistic abstraction, and therefore bases its philosophy upon objective premises whose value is relative and changeable. Upon such a shifting foundation no stable social order can be built. The materialist believes that the scientist has disproved the existence of the spirit when he has merely proved the literal falsity of legends whose symbolism is no longer significant. He has not invalidated one iota the absolute truth of which they were once the symbol, for although he is able to answer the question—How? with ever increasing confidence, he does not and cannot answer Why. On the contrary the very complexity and subtlety of the processes which man has discovered in the course of researches in evolution and natural law disprove the materialist argument.

These relative and changeable values upon which materialism rests are of their very nature incapable of providing any common basis or purpose upon which the life of the individual or of the community can be built. As a result modern life is the antithesis of the mediæval conception of wholeness and self-sufficiency. Our thought is confused and disassociated, we cannot distinguish between ends and means, cause and effect. The individual's work, home life and leisure are arbitrarily divided; there is no association between them nor can they be satisfactorily related to the life of the community; because the course of our lives is directed to a material end we are necessarily predatory, and our concern to extract the maximum monetary reward for the minimum of effort and responsibility isolates us from our fellows to make us insignificant units of a herd instead of responsible members of a community. All the propaganda resources of popular press and radio, not only in this country but all over the world, are increasingly engaged to disprove the existence of this tragic separatism by investing national and even international groups with a unity of purpose and belief which does not exist in fact, and is, therefore, a pure abstraction. Each individual knows in his heart if he be true to himself that he is no longer a responsible member of a community, but an irresponsible unit in a world where the race is to the swift and the battle to the strong.

Great responsibility in the affairs of men means great authority, and such authority spells power which corrupts. This is a problem with which the human race has always been faced. It can only be solved if human authority acknowledges laws greater than itself, and if responsibility is distributed as widely as possible in the community. We have not only based our civilization upon relative abstractions, but attempted to solve this problem of power, not by the distribution of responsibility, but by its dissolution in collectivism with disastrous results.

There could be no better proof of the existence of this separatism in our society than that afforded by modern sociological attempts to overcome it, notably the developing technique called "mass-observation" whereby earnest intellectuals penetrate the (to them) unexplored jungles of factory and public house, as though on safari in darkest Africa. Their humourless reports, couched in that pseudo-scientific jargon beloved of the modernist, treat of individual human behaviour as though it was a chemical or physical phenomenon. Such attempts to reduce humanity to the terms of a laboratory experiment owe their origin to the system of " Market Research " developed by the great industrialists as a logical corollary to their nation-wide advertising campaigns. Their aim being to dupe the public in order to secure a profitable market for their shoddy mass-produced goods, it was obviously advantageous, on the pretext of " studying public need ", to discover the psychological effects of that advertising. When the rulers of a nation adopt such methods as a basis for propaganda, " mass-observation " creates the mass-mind, furthermore it may all too readily assume the sinister office of a secret police force.

One of the most disastrous results of separatism is the completely uncritical attitude of the majority of intellectuals towards the scientist, the technician and the specialist. No edict of oracle or high priest in the ancient world could have been accepted so unquestionably by the whole people as is the portentous pronouncement of the specialist to-day. Consequently the modernist regards the impact of new technocratic developments upon society as being, like sunrise, inevitable, even though he may perceive their potential danger.

While such a state of affairs exists no improvement of communications or world-wide propaganda about " solidarity " and " the brotherhood of man ' can restore that sense of responsibility without which there can be no social stability. So long as our outlook continues to be materialistic and predatory, so long will our lives continue to be governed and determined by forces beyond our personal control. We shall continue to perform work whose purpose and ultimate effect upon humanity we do not know and are powerless to influence, and we shall be controlled in every walk of life by complicated machinery the construction and operation of which we are ignorant. The scientific " Progressivist " argues that this growing complexity is the logical product of greater knowledge, and that any social evils which result from it are merely due to our failure to adapt ourselves efficiently and promptly to changing conditions. This view is typical of modern thought in its short-sighted confusion of cause with effect, for the complexity of modern life is not the product of growth but of disintegration; we are witnessing, not the laborious construction of a new world, but the ruin of a civilization which, like a tower built on a quicksand sinks faster than we can add brick to brick.

> "Things fall apart, the centre cannot hold,
> The falcon cannot hear the falconer."

Unless we resolve to build anew upon sound foundations we can only prevent our crazy edifice from collapsing into chaos by reinforcing it with the steel of authoritarian government. The more top-heavy the structure the wider the cracks and the stronger the steel. The function of such a government, whether it profess itself to be Democratic, Aristocratic, Fascist, Socialist or Communist, can only be to exercise a rigid control over self-interest to secure economic justice. Since it is founded upon material and relative values it can only speak in these terms and cannot therefore invoke moral principles of absolute truth. Any attempt to do so by its spokesmen must be hypocritical and regarded with cynicism. To enforce its will it must encourage national and racial egotism by means of education and propaganda, and be prepared to impose the necessary uniformity and docility by means of elaborate bureaucratic and technocratic machinery. Such a form of government confers upon those who impose it dangerous and irresponsible power which no minority can be qualified to wield.

We are therefore faced with the choice between chaos or the slave State unless we are prepared to acknowledge our error with due humility and, by re-discovering spiritual principles as the only sound basis of living, restore the lost dignity of individual responsibility and self-sufficiency.

All the conflicting political ideologies which clamour for the public ear and contribute to the bewildering babel of the popular press are founded upon the material expediency of relative truth and are therefore incapable of bringing about such a reformation. Even in the support of their limited ends there exists no common purpose within their ranks, for the majority of their supporters are ignorant of political theory, while the intellectual party caucus has no practical experience of the everyday life of the rank and file. While the urban philosophers pursue their sterile theories, Jack joins the Communist party because he envies his master and believes that communism will give him a larger share of the world's wealth, being under the erroneous impression that in Russia Utopia has already arrived. By the same token his master supports the reactionaries because they safeguard his interests against communism. Thus the potency and popularity of all these ideological products of modern thought depends on the prospects of material prosperity which they offer to one class or another, and their most passionate partisans are those most concerned with their own self-interest. This sorry truth has never been better or more eloquently expressed than by the late W. B. Yeats. In two brief lines he does not merely write a fitting conclusion to this chapter, but summarizes for posterity the tragedy of our generation:

> "The best lack all conviction, while the worst
> Are full of passionate sincerity."

Chapter VIII

THE TWILIGHT OF THE ARTS

So low have the arts fallen and so small is the part they play in the life of urban man that to devote a lengthy chapter to art and the artist in a social survey so broad as this may seem to be irrelevant. This is not so. For the true artist is concerned with the expression in relative terms of the aspects of absolute truth which he perceives through his æsthetic sense. This being so his conception of work is necessarily qualitative; he is, in fact, a craftsman since the eloquence of his expression is dependent upon his sympathy with the medium in which he works and on his skill in its use. His work differs only in degree from that of the craftsman since it achieves consciously and deliberately what the other achieves intuitively. Because both are the flowers of the freedom of human ability their work manifests an aspect of truth, and because it is true it is beautiful since beauty in any form is simply the revelation of truth. If, therefore, we except the craftsman whom we have almost exterminated, the artist remains the only custodian of values which our civilization has discarded, and on this account the vicissitudes of art from the mediæval period to the present day are far from irrelevant.

The artist of to-day, if he remain true to himself and to his art and does not fall a victim to one of the many pitfalls and abstractions which lie in his path, must necessarily be a rebel; an intractable individualist in a world increasingly hostile to individuality. At the worst he will be persecuted, at best regarded with ridicule by those who dub him "high-brow", "crank", "idealist" or "escapist", and cannot comprehend that it is he who is aware of reality and that it is they who escape. For the artist is a rebel against the mind which will only accept those things which it believes to have been proven by scientific or pseudo-scientific jargon, and which attempts to reduce transcendent greatness to the narrow compass of objective determinism. Art does not seek to explain or to reduce in this way; it is concerned to express the inexplicable in a language of symbol and image. The artist evolves this language in the loneliness of his moments of deepest perception because he feels an instinctive desire to express, and so to share, some inkling of that eternal association of beauty with sorrow which lies at the heart of all things, and of which all men's minds are dimly aware. Art is an acceptance of life, not an escape. It acknowledges a reality greater than ourselves which men can approach only by giving of their best in their brief stay. It perceives that, just as the flower's beauty of colour and fragrance lies in the knowledge that it must wither and fall only to be renewed by another spring, so perfection of the human life is dependent upon, and yet alone can transcend, mortality. The great artist is he who perceives and expresses most clearly the beauty of this tragic and eternal association of opposites; realizes that life would lose all sweetness if it were immortal, since without sorrow there can be no joy, without death no renewal, without immanence no transcendance. In this way the perspective of his vision is altered so that external objects and events are translated from their relative and transient environment to be judged in a context absolute and changeless. This deeper vision which we call imagination or inspiration is a faculty of the æsthetic sense ; to express his vision the artist makes use of reality as material to create symbols which outlive their perishable substance. Thus a fleeting pattern of shadows, a chance word or gesture, events that exist objectively only for a moment of recorded time, may conjure in the imagination some suggestion of eternal truth and beauty. If the artist is able eloquently to translate such an experience into the language of his art, then the moment will live immortally for the enrichment of other minds as an incantation of words full of a far off suggestion that transcends their literal meaning, as a tapestry of sound intricately woven, or as a subtle grouping of colour or of form.

Such translations represent the highest achievement of which man is capable. Though they are the works of individual genius rare in any age they are also the product of the civilization with which they are contemporary, because the artist's raw material is the life and thought of his time. If that life is founded upon principles of truth, and therefore presents an organic harmony of knowledge and wisdom, then all is pliant clay in the artist's hands and his art achieves a universal eloquence. For life is lived in accordance with his own principles and he is both understanding and understood. If, on the other hand, the life of that civilization is devoted to materialistic and relative ends it is false and therefore provides material which the artist cannot use. His task becomes increasingly difficult. His lot is that of some master mason searching amid the shale of a worked-out quarry for a solid bed to which he can set his chisel. His search and his work can only meet with partial success; either he works upon a crumbling stone that flakes and cannot fulfil his design, or he despairs and, retiring to his workship, labours at some old stone long cherished from the past. Whichever course he takes his art has no universal quality; he may speak with a small voice which will please many for an hour, or cry aloud of lost greatness in speech unintelligible to the multitude.

In the Middle Ages that harmony of which we have spoken was to a limited extent achieved, and it was manifested in the life of the mediæval village community. This was in the truest sense of the word an organic or functional life since it called forth the maximum of ability and self-sufficiency from the individual in co-operation with natural law. Because man's life was built in accordance with that law his object was not to "conquer" natural forces and processes, but to acknowledge them and adapt himself to them. For this reason his life achieved that "wholeness" in which it is impossible to separate one function from another. Consequently, there was no such thing as art in the sense in which we speak to-day, and yet in another sense all men were artists. The village mason who glorified church, manor, barn and cottage alike, created what we now consider works of art, yet he was no self-conscious artist striving for effect; he was a craftsman, and, because wisdom and truth drove his chisel, beauty followed it as irrevocably as the flower breaks from the bud. *Laborare est orare*—an intuitive allegiance to this principle informed every aspect of life and work, a fact which is evident to-day in those monuments which the age has left. It found its reflection also in speech and song full of natural poetry, and it was the pursuit of this lost eloquence that led J. M. Synge to forsake Paris for the Aran Islands.

Perhaps it is the influence of the Irish writers of the so-called "Celtic Twilight" period which has given birth to the idea that natural poetry is a special gift of the Celtic people. In fact, though some may excel others, it is a gift possessed by all peasant peoples, the fruit, in other words, not of one race, but of one way of life. It has almost vanished in England because that way of life has been destroyed, and not because our forebears possessed no natural gift. It is only within the last half century that the last traces of that art have vanished from life to survive only on the printed page. For example, our fathers might have heard "The One O" sung at Harvest Home. This song discovers a well of memory whose profound depths we can no longer plumb; its strange cumulative refrain swelling to a finality that is like an incantation:—

"A song I will sing you.
What may your song be?
I will sing the twelve O.
What may your twelve be?
Twelve are the twelve Apostles,
Eleven are the eleven Evangelists,
Ten are the ten Commandments,
Nine are the bright shiners,
Eight are the bold rangers,
Seven are seven stars in the sky,
Six are the proud walkers,
Five are the thimbles in my bosom,
Four are the Gospel preachers,
Three, three O are rivals,
Two, two are lily-white boys
Clothèd all in green O,
And when a man is dead and gone,[1]
He's no more to be seen O."

The spread of the written word coupled with the new intellectual freedom which accompanied the Dissolution and Reformation gave birth to a great flowering of deliberate recorded art. Art acquired its capital A, and soon a further distinction was to be drawn between the Artist and the craftsman, between fine and applied art. The first fruits of this "Renaissance" constitute some of the greatest achievements of Western European civilization. The literary achievements of the Elizabethans were universal in quality and of an immortal stature because the artist was perfectly in harmony with the life and thought of his time,

[1] All these ancient songs vary from region to region, and I have seen several different versions of this one, sometimes under different titles. This version comes from Alfred Williams' collection, *Folk Songs of the Upper Thames*. He obtained it from William Jefferies, of Longcot, and William Wise, of Alvescot, Oxon.

with its principles and beliefs which were still mediæval, and could thus employ every aspect of that life as material for his art. Conversely, because his principles were not at variance with those of the people, but were developed from them, his imaginative conception could be readily conveyed to them, their own language affording him a perfect vehicle of expression. Consider, now, this work of the Renaissance:

> "When I consider, everything that grows
> Holds in perfection but a little moment,
> That this huge stage presenteth nought but shows
> Whereon the stars in secret influence comment;
> When I perceive that men as plants increase,
> Cheered and checked even by the self-same sky,
> Vaunt in their youthful sap, at height decrease,
> And wear their brave state out of memory;
> Then the conceit of this inconstant stay
> Sets you most rich in youth before my sight
> Where wasteful time debateth with decay,
> To change your day of youth to sullied night;
> And all in war with time for love of you,
> As he takes from you, I engraft you new."[1]

This is great love poetry, but it is also something more. Couched in the language of the people, it employs the same natural allegories to express that tragic view of man as part of an eternal process of decay and renewal which no one who is in close touch with the realities of the natural world can escape. All this the poem shares with the instinctive art of the people. Its greatness lies in the deliberate employment of this traditional material by an imagination of great creative power. It is a difference of degree only, a flower that has sprung from deep and ancient roots. In the poem's final triumphant line the potential greatness of the human spirit, the ability of man's creative imagination to transcend the flux of time, is magnificently affirmed. Yet, and this is the most important point of all, there is no intellectual pride in this triumph. Here there is no arrogant rejection of life; the great conclusion follows a tacit acceptance of life's simple realities. For Shakespeare realizes that the reason why his love appears to him so rich in youth, is the knowledge that her perfection is, like all things, transient; that it can "hold in perfection but a little moment."

[1] Sonnet XV.

Because it was the full flower of an age that had passed and not the root of the new age which had been born, the art of the Renaissance soon withered and lost its universal quality. This decline became inevitable so soon as deliberate art divorced itself from the life of the people on whose traditions it had been founded and no longer spoke in their language. While the instinctive art of the people survived to die a lingering death as the way of life which inspired it perished, the deliberate artist became the lackey of the new aristocracy whose principles we have already examined. This nobility of counsellors, landowners and merchants true to their philosophy, abandoned the art of their people and concentrated upon, as more fitting to their high estate, the art of Greece and Rome, which was the stimulus of the Renaissance.

While no one would deny the greatness of the Classical Tradition or the success with which it was re-interpreted in the seventeenth and eighteenth centuries, it should be obvious that no art that has been brought to perfection by a vanished civilization can ever be fully assimilated by a people of whose life it is not the logical expression. Because it plays no organic part in that life, any greatness it may achieve is necessarily limited and transient; it is subject to change at the whim of fashion, a process which leads to emasculation rather than to growth. So it was with Classicism in England. Though its influence was far-reaching, it was essentially the exclusive "culture" of a class. Like religion, the arts withdrew from the business of living; they forsook the field and the market place.

The course of this withdrawal is not difficult to trace. The Elizabethans paid little more than lip-service to the new mode. Shakespeare may have set the scene of his *Midsummer Night's Dream* in Greece, but his "wood near Athens" is his own forest of Arden, while *Pyramus and Thisbe*—that play within a play, is a piece of rural mummery, such as he must often have witnessed, played by his own countrymen of Warwickshire, and descended, no doubt, from the mediæval dramas of the Guilds. The most sublime poetry consorts amicably with the colloquial phrase; even his immortals are steeped in the lore of rural England, as when Titania speaks of the disasters of a wet harvest:

> "The ox hath therefore stretched her yoke in vain,
> The ploughman lost his sweat; and the green corn
> Hath rotted ere his youth attain'd a beard."

Contrast this with Milton. Here, too, is great poetry, but it is poetry written for an educated class that is forgetting its roots in the soil and breaking its links with the Commons. The wild wood which harbours Comus and his crew lacks roots; classic myth and allusion overlay the native idiom, and Sabrina is no true spirit of Severn and of the Western March, but nymph of a river which flows only in the Paradise of the poet's imagination. That imagination may reach to the stars, but Milton's feet have left the earth. With this fatal severance comes intellectual pride, and in *Paradise Lost* it is Satan who holds the stage despite the poet's moral argument.

From the grandeur of Milton it is but a short step to the prolixity of Dryden, who attempted to re-write Shakespeare in the fashionable mode with small success, and so the classic glory waned. As a conventional form it lingered on until the Victorian age, changing its mood at the whim of fashion. The solemnity of the Commonwealth gave place to the artificial gaiety of nymph and muse at the Restoration, when in extravagance and licence a self-conscious aristocracy sought in vain to recapture the lost spirit of a " Merry England " that had gone beyond recall. This mood in turn was succeeded by the polished and deliberately artificial elegance and wit of the Augustan age, a brilliance which was finally overwhelmed in the closing years of the eighteenth century by a second wave of Puritanism for which the rise to power of the industrialists, with their confused conception of morality, was responsible. Under their ægis the classic tradition spent its last breath as the windy instrument of fulsome praise or pious moralising to be replaced in fashionable favour by the Romantic Movement.

The Romantics revolted against the deliberate order of Classicism, and their reaction took the form of a return to nature for inspiration. Yet they did not attempt, nor could they have achieved any reintegration of art with life, but merely substituted for one set of conventions another no less artificial. To them " Nature ", with a capital N, was an Awful Wilderness or Smiling Mead, a ruin Picturesque, the surviving countryman, ignorant and rightly suspicious of these urban conceits, a Rustic Boor.

By this time, however, the serious artist was no longer so beholden to a wealthy class who, unlike the Augustan Patricians, did not consider that patronage of the Arts was a *sine qua non.* Consequently, he became an individualist and a rebel like Percy

Bysshe Shelley. This liberation of the artist from the conventions dictated by aristocratic patronage, although it brought about no immediate rapprochement between art and life, at least gave the artist greater freedom, and it was this freedom, coupled with the manifest degradation of the applied arts, that gave rise to the ferment and conflict of ideas which distinguished the end of the last century.

At this point, it is necessary before proceeding further, briefly to review the divergent development of the applied arts from the Renaissance onwards. Prior to that date, as we have seen, these arts, of which the most manifest is architecture, were the responsibility of rural craftsmen skilled in the use of local materials. It is significant that, although much of their work still survives in all its beauty, they remain anonymous, being the product of many hands working in mutual responsibility and harmony. It was not until the latter end of the sixteenth century that the first signs of specialization and division of responsibility appeared and we begin to attribute certain buildings to individuals. These were "designer-surveyors" appointed by the builders to superintend the work, men such as Shute, Lyminge and Thorpe. In a few years time specialization was carried a step further and the function of design was entirely separated from the practical work of construction. Inigo Jones, the son of a clothmaker, persuaded Lord Pembroke to send him to Italy, where he absorbed the principles of Palladio; his return marks the beginning of the age of the Architect and the general introduction of Classicism to the applied arts.

For the gracious architecture of the ensuing period we are inclined to give too much credit to the architect and too little to the craftsmen responsible for translating his design into concrete terms. The former was a product of the new age, an exponent of a new school of thought, the latter was the inheritor of principles and traditions which were quickly passing away. The diversity of architecture and of other forms of applied art during this period was the result of this interaction between old and new. In remote districts and in humbler forms the traditional work of the craftsmen remained dominant, while in and around the greater towns the architect held undisputed sway, the craftsman being merely a docile executant. In the former work we can detect that impoverishment of ideas inevitable when a tradition survives in a hostile age, in the latter an arrogance, a concern for display, and a contempt for local environment and materials which epitomizes the philosophy of the new urban civilization that was being born. The happiest products of this transitional period are those in which the principles of the architects were translated and modified by the craftsmen to suit their traditional methods and local materials, for in this way the noble classical proportions were merged into, and infused with new life, the organic principles which the craftsman intuitively perpetuated. Without this close harmony of common principle between design and execution, the product conveys the sense of something lacking, no matter how well proportioned and impressive it may be. In this regard it is of profound significance to note that the mediæval builder's crowning achievement was the classical architect's most conspicuous failure; his churches, for all their costly magnificence and meticulous sense of proportion, being so many mausoleums of a dead faith. This is well illustrated in William Woty's poem "Church Langton," in which he compares Westminster Abbey with St. Paul's:

> "Lengthening ayles, and windows that impart
> A gloomy steady light to cheer the heart,
> Such as affects the soul, and which I see
> With joy, celestial Westminster! in thee.
> Not like Saint Paul's, beneath whose ample dome
> No thought arises of the life to come.
> For, tho' superb, not solemn is the place
> The mind but wanders o'er the distant space,
> Where, 'stead of thinking on their God, most men
> Forget His presence to remember Wren."

As the eighteenth century wore on so the influence of urban taste expressed by architect and designer waxed while that of the craftsman waned. The colossal erections with which Vanbrugh satisfied the grandiloquence of his aristocratic patrons were the portents of this change. Improved methods of transport for both men and materials led to an eclipse of the craftsman's regional influence, and the urban style was imposed upon the country to an increasing extent. At the same time the cheap manufactured goods of the rapidly developing industrial areas began to oust rural and regional craftsmen in all the other spheres of applied art. For the most part these goods were cheap, badly reproduced copies of the prevailing urban fashion, damaging alike to the integrity of the art which they copied and to the critical faculty of those to whom they were supplied. When the style upon which these reproductions were based itself became degenerate, and mechanical methods of mass reproduction were perfected, the debasement of applied art was almost complete. The turning point was reached at the close of the century when the great age of the classical architect enjoyed a final flowering in a period of sophisticated elegance and grace for which the influence of the brothers Adam were largely responsible. Thereafter, architect and designer pandered to an ever increasing extent to the pretentious whims of the *nouveau riche* commercial class which was rapidly becoming dominant in national affairs. It is unnecessary to do more than mention the result of this unholy alliance, or to devote any space to such travesties as the Victorian Gothic Revival. Our cities contain plenty of examples in bogus Classical, Gothic, Scottish Baronial, and Baroque or their admixture which afford concrete evidence more convincing than any words of the depths to which the applied arts sunk in the last century.

It was out of the new spirit of rebellion which, characterized the exponents of the fine arts towards the close of the Victorian Age that there sprung the first attempt to bridge the gulf of three centuries, to reconcile art with life, and by so doing to rescue the applied arts from their slough of despond. Broadly speaking the artists of this period grouped themselves into two opposing schools, the one Realist, the other Traditionalist. The Realists may be so-called because their artistic philosophy was based upon an acceptance of modern life as they found it. They did not seek to change it or to establish any principles, but rather to distil through the medium of their art, every emotion and experience which life could bring them. As a result there developed what has been called the *fin de siècle* renaissance of the fine arts, an art of the city in which artificiality and separatism was carried to its ultimate and most rarified conclusion, by the writers and artists of the "decadence". These men professed an intellectual dandyism which they evolved in part from a misconception of the austere philosophy of sensation propounded by Walter Pater, and to a greater extent from the influence of French Symbolists Baudelaire, Verlaine, Rimbaud and Mallarmé.

In their view art was upon a plane above life and nature, while the artist was a kind of secular high-priest of sensation and emotion, sampling good and evil alike, and ever becoming more eclectic as his coarser appetites were satisfied. The drawings of Beardsley and the equally exotic writing of Wilde are typical expressions of this extreme form of intellectual æstheticism which, though it professed an acceptance of life was, in fact, an escape, since it would only accept on the condition of withdrawal into an abstract world of fantasy. "As for living", declared Villiers de L'Isle Adam, "our servants will do that for us". "All bad art", wrote Oscar Wilde, "comes from returning to life and nature, and elevating them into ideals. Life and nature may sometimes be used as part of art's rough material, but before they are of any real service to art they must be translated into artistic conventions".[1] This *fin de siècle* tendency of the decadents was thus a form of what we now term escapism, a withdrawal of the artist into an ivory tower to which he would admit life only upon his own terms. In this exalted seclusion he tended to dabble in brands of esoteric mysticism which an earlier and wiser age would have called satanism.

The artists of the traditionalist school, on the other hand, though equally conscious of the separation of art from life, sought no escape from reality but endeavoured to bring about a reunion with life through the medium of applied art. After the lapse of three centuries they made a gallant but abortive attempt in the face of a world drugged with material prosperity to bring artist and craftsman together once more by asserting the qualitative, as opposed to the quantitative conception of work. The history of this movement and the reason for its failure is thus of great significance and is worth considering in some detail.

Its inception was largely due to the writing of Ruskin, notably his *Stones of Venice,* wherein he declared that art was the expression of man's joy in his work, and that the aim of all artists should be "the hallowing of labour by art". Although, as Chesterton pointed out, Ruskin knew more of the carvings of a mediæval church than of the faith which inspired them, and was therefore himself an example of the separation of art from life, this teaching of his re-affirmed an important though limited truth, and it did not fail to bear fruit. It led to the inception of the Arts and Crafts Society, and to the foundation of the Arts and Crafts Exhibition Society by Walter Crane in 1886. Meanwhile these movements were accompanied in the sphere of the fine arts

[1] From Wilde's *Decay of Lying,* which might have been sub-titled *A Defence of the Abstract.*

by the Pre-Raphaelites, who, as their title denotes, sought to pick up the lost threads of the art that had perished at the Renaissance. As must always be the case with any revival of the past in a hostile and fundamentally different age, the art of the Pre-Raphaelites was even further removed from life than that of the Decadents and the later Impressionists, for whereas the latter built their dream world out of the present, the former fashioned for themselves a romantic but archaic life out of mediæval legend.

It was not one of the original Pre-Raphaelite Brotherhood, but a late-comer to the movement, William Morris, who realised that the principles of art could only be restored to their place in life by a far-reaching change in the structure of society. He saw, too, that such a change must imply the downfall of the very class upon which, being a "luxury trade", the existence of both fine and applied art movements depended; "The arts have got to die," he wrote, "before they can be born again."[1] With the exception of Tolstoy he perceived perhaps more clearly than any man of his generation the depths of spiritual degradation and economic slavery to which the life and work of the common man had sunk. Yet in a period of unexampled material prosperity and expansion his voice fell on deaf ears, though many of his sayings remain as true to-day as when they were spoken. He devoted the whole of his life to the belief that only the application of the principles of art to life could break the vicious circle of that utilitarianism which he defined as "the reckless waste of life in pursuit of the means of life".

He sought to implement his ideals in two ways, by direct political action, and by practical example. In 1883 he joined the Democratic Federation, the only Socialist organization then in being, while to the latter end, as an accomplished craftsman, he became the leading spirit of the Arts and Crafts Exhibition Society. Like that of so many idealists who carry their sincerity into material affairs his political career was brief and tragic. The associations of his art and the sheltered circles in which he had moved had not taught him what to expect. Perhaps he fancied that the working class of Victorian London were so many craftsmen with ideals like his own who only awaited his coming to head their march to the promised land. But the gulf he sought so lightly to leap yawned infinitely wide. He found himself among men moved by the same motives of self-interest as their

[1] *Hopes and Fears for Art*—William Morris.

rulers, men who had adopted Socialism as a means of political or economic advancement, men who mouthed fiery phrases, but slunk away submissively when their rulers cracked the whip. To the ears of these men his talk of art and life was as unintelligible as double Dutch. Morris's inauspicious excursion into real life virtually came to an end on "Bloody Sunday" 1887, when a demonstration march on Trafalgar Square was forcibly dispersed by police and Life Guards. Thereafter he retired to his manor at Kelmscot and his London house of the same name, where he devoted himself to preaching his ideals through the example of his craftsmanship.

The Arts and Crafts movement to which Morris contributed so much exerted an influence more evil than good, the effect of which is still apparent to-day. Because it proved powerless to affect any social change it necessarily remained artificial and apart from life. It put the cart before the horse by attempting to bring about an artistic revival by imposing an archaic standard upon the existing system, whereas a genuine renaissance can only come into being as the natural product of a change within the system. The history of the movement must therefore stand as a warning for all time to those who seek to impose the *results* of the past upon the present. Necessarily it remained a luxury trade, but the popularity of its products soon attracted the notice of the enterprising manufacturer who proceeded to flood the market with debased machine-made replicas. The movement also fostered that false and empty sentimentality and nostalgia about the past which was to bear horrid fruit in our own times as an outbreak of Olde Worlde Tea-Barns, Show Villages, Anciente Hostelries, Picturesque Tudor Residences and Crafte Shoppes, which rival the worst extravagances of the Victorians. It led to the wholesale exploitation of the things of the past by a people totally ignorant of the principles that informed the way of life of which they were the natural product. The few remaining countrymen found themselves evicted from their homes into brick-box council houses to make room for week-end rustics.

Nevertheless the work of William Morris was not wholly set at naught. The contrast between his own work in the field of applied art and that of the tasteless Victorian manufacturer and decorator was too great to be ignored. In certain respects, notably in the design of printed fabrics and wallpapers, and in the work of book production and typography, he inaugurated a process of progressive improvement. Furthermore, Morris passed on his ideals to younger craftsmen, such as Ernest Gimson and Sydney Barnsley, who have together been responsible for some of the finest furniture that our impoverished century has so far produced.

So ended the first and only large-scale attempt to reconcile art with life. Its heyday was brief, its experiment in the applied arts soon commercially exploited, while its sham-mediæval influence on the fine arts vanished like the pipe-dream which in fact it was. Pre-Raphaelitism having spent its force, it was left to the Impressionists and Post-Impressionists of the opposing school to uphold the fine arts in the twentieth century. As the young century drew on, however, serious artists (and by the term serious we exclude all those who succumbed to the clamour of popular applause or sought to advertise themselves by the extravagance of conceit), tended to take their place in one of two schools of thought. Upon the one hand were those who still cherished the idea of the artist as a kind of high priest standing aloof and dispassionate above life and, while not professing to ignore its topical manifestation, accepting it only upon his own terms and in so far as it could be translated into a conventional language and idiom. Upon the other hand were ranged the younger artists who had grown up in an age of war and social disintegration and had never experienced the ease and security of the previous century. They felt that the measured beauty, the roses and lilies of the older forms of artistic expression afforded them no adequate vehicle in which to express their acute perception of the problems of the complex, crumbling, and distraught material civilization in which they found themselves. Theirs was a new world of machine cities which could not be expressed in the language of the old. How could they voice their convictions except by using the language of this world and by evolving new forms and new symbolism? They must sing of the machine, the pylon and the by-pass road, and in this way create a modern art language intelligible to modern men. This idea has failed as it is bound to fail for two closely associated reasons; firstly, because the material manifestations of the modern world are the product of relative, limited, and therefore false values, and are consequently ill-fitted to reveal universal and absolute truth no matter how skilfully they may be used; secondly, because the complexity of modern life which is the result of separatism and specialization renders impossible any common understanding between the artist and the specialist in whose language he is trying to speak. The artist may employ technical symbolism to his own satisfaction and to that of his own intellectual clique, but to the man in the street, or to the specialist in the particular branch of technology he has chosen, its significance will be lost even if it is not ludicrous nonsense which unfortunately is more often the case. The only poet who has so far succeeded in using the language of the machine effectively was Kipling. He could do so because he was only concerned to express in his verse relative values which were in perfect accord with the philosophy which developed the machine. Therefore he has the ear of the technician when he writes:

> " The crank-throws give the double-bass, the feed pump sobs and heaves,
> An' now the main eccentrics start their quarrel on the sheaves;
> Her time, her own appointed time the rocking link head bides,
> Till—hear that note?—the rod's return wings glimmerin' Through the guides."

The modern poet, on the other hand, when he tries to use the same language to convey deep imaginative feeling, fails:

> " The great cranks rise and fall, repeat,
> The great cranks plod with their Assyrian feet
> To match the monotonous energy of the sea."

These lines may be effective in their appeal to the literary " high-brow ", but the comment they would evoke from a sweating greaser is best left out of print. As though to emphasize the failure of this school of artists to assimilate technics, another poet of the modern school, commenting on these lines, speaks of the " turbine engine ". As steamships are either powered by reciprocating engines or by turbines which do not employ the principle of the crank, his comment still further reduces the practical validity of the lines he quotes.

To avoid falling into these innumerable technical pitfalls, and at the same time to attempt to convey his message to the " man-in-the-street " the artist, having eschewed also the traditional language of art, can only work between the narrow limits imposed by the current level of popular entertainment and compulsory education. To attempt to create a work of art with this material is like trying to make pottery without clay, and all that the sincere artist can do is to become teacher and propagandist. Only in this way can the artist come to terms with reality at the risk of losing his own soul, for the poet of the so-called Pylon School, no less than the older exponent of traditional forms, writes for the clique and coterie.

It is significant that W. B. Yeats, whom it is probable that posterity will regard as the dominant figure in the art of our period, remained faithful to traditional form. With consummate genius he proved that traditional language and symbolism could yet remain flexible and adaptable to the changing mood and thought of modern life. In his early years he was profoundly influenced by the Symbolists, and his early poems of Celtic legend were full of that elaborate and cloudy imagery, that over-charged verbal magic, which characterized the school of Pater, Baudelaire, Mallarmé and Villiers de L'Isle Adam. In later life, however, his art emerged from these " dim, heavy veils " of Celtic Twilight to assume a new and austere grandeur. He gradually shed superfluous words as a man in training for some marathon sheds superfluous flesh until, in the apparently simple ballads of ancient form with their recurrent refrains which comprised much of his most mature work, every word and phrase was charged with dynamic vitality like the taut muscles of the athlete. With this style, stripped of all " purple passages ", he challenged the modern school upon their own ground. He proved, moreover, that the traditional forms which he used were, when moulded by his genius, capable of greater eloquence than the modernist symbolism of the younger artists. Yet the difference between Yeats and the Moderns is not merely one of style, but of philosophical conception, a difference profoundly significant. For whereas the majority of the Moderns accepted the tradition of humanism, of " liberty, equality and fraternity ", and therefore espoused the cause of collectivism, Yeats remained an implacable individualist. He perceived that the socialist aim of equality could only be achieved within the framework of modern civilization by the extinction of all those individual qualities that are the inspiration of all art, and which alone can make men great. He was deeply conscious of that ebb of the spirit which coincided with the flow of reason from the

Renaissance onwards, and, in the early essay upon Edmund Spenser, he wrote:[1]

"When Spenser was buried at Westminster Abbey many poets read verses in his praise, and then threw their verses and the pens that had written them into his tomb. Like him they belonged, for all the moral zeal that was gathering like a London fog, to that indolent, demonstrative Merry England that was about to pass away. Men still wept when they were moved, still dressed themselves in joyous colours, and spoke with many gestures. Thoughts and qualities sometimes come to their perfect expression when they are about to pass away, and Merry England was dying in plays, and in poems, and in strange adventurous men. If one of those poets who threw his copy of verses into the earth that was about to close over his master were to come alive again, he would find some shadow of the life he knew, though not the art he knew, among young men in Paris, and would think that his true country. If he came to England he would find nothing there but the triumph of the Puritan and the merchant—those enemies he had feared and hated—and he would weep perhaps, in that womanish way of his, to think that so much greatness had been, not as he had hoped, the dawn, but the sunset of a people".

Yet although Yeats rejected materialism he could not, perhaps because of his intellectual pride, accept the religious view of life. Consequently he took the only possible middle course by exploring the realms of magic and esoteric philosophy, researches which culminated in the writing of *A Vision* where, by means of that elaborate system of geometrical and lunar symbols which deeply influenced his later poetry, he expounded his theory of a cyclical system controlled by occult forces. According to this theory it seems clear that he envisaged the breakdown of our present civilization and the emergence from the ensuing "dark age" of another phase similar to the mediæval period. These findings were very similar to those of Spengler, although, at the time *A Vision* was first written, Yeats had not read Spengler's work. This was not altogether a coincidence, for, however widely application and conclusion may vary, any philosophy which assumes the existence of superhuman forces, while it rejects the religious doctrine of free-will, must accept the idea of pre-destination. Consequently it remains as deterministic as the philosophy of materialism by simply making man the sport of occult rather than material

[1] Collected in *The Cutting of an Agate*.

forces. In fact, Yeats, like Hardy, was a fatalist. Man is equally incapable of influencing the course of events for good or ill whether he is controlled by the great wheels and gyres of *A Vision*, by the dispassionate spirits of *The Dynasts*, or by Shaw's "Life Force."

We may not agree with Yeats' philosophy and prefer to believe that it is within our power to undo the errors of the past three centuries, but we cannot deny that Yeats possessed a most profound perception of our social maladies. Dismissing modern Utopian notions of human equality as a "formless, spawning fury", he realized that society must always consist of rulers and ruled, and consequently he was most concerned to determine who was best qualified to rule. He believed that the artist, if not himself the ruler, should be the power behind the throne, and it was in this belief that he concerned himself with Irish politics. Owing to a facile misconception of these views, modernists have accused Yeats of Fascist tendencies, but Yeats, with his infinite sympathy for individual personality, was the last person to advocate a servile state controlled by a bureaucracy of technicians, demagogues or professional thugs.

Despite his political activities, and despite the influence of J. M. Synge which led him to "sing the peasantry", it is doubtful whether Yeats ever ceased to think of the artist as a special kind of person as opposed to Eric Gill's view of every man as a special kind of artist. Consequently, whereas William Morris had sought to restore the principles of art to their place in the everyday life and work of the common man, Yeats envisaged no such re-union; his view was aristocratic, his ideal a strong highly cultured heirarchy of wisdom and wealth. The Abbey Theatre, and the plays which Yeats wrote for that stage may represent a great attempt to restore imaginative drama by re-uniting author, producer, actor and audience in close collaboration, but it is significant that Yeats eventually abandoned this attempt in favour of plays written for the private salon rather than for the public stage.

Yeats' great poetry will prove immortal because its greatness lies, not in philosophic argument, but in its revelation of universal and eternal truths. Yet he could not bring the artist's conception of reality one step nearer to the common man. He could speak only to a minority, and even of these only a few could follow the soaring flight of his vision. When he died in 1939, on the eve of the greatest war in history, he left the arts as he had found them, not the voice of a great people, but the æsthetic luxuries of a minority, preserved under the collective title of "Culture" like so many rare plants in a hot-house. When so preserved, the arts remain the concern of the specialist, their current forms reflecting the exclusive idiom of a clique or a shallow Alexandrianism, their past greatness an embalmed splendour.

Like Isolt of Brittany, the true artist cries:

> ". . . . I am not one
> Who must have everything, yet I must have
> My dreams if I must live, for they are mine.
> Wisdom is not one word and then another
> Till words are like dry leaves under a tree;
> Wisdom is like a dawn that comes up slowly
> Out of an unknown ocean."[1]

Let him be thankful for the visions of these dreams, and for any wisdom they may bring him, for they are a priceless heritage. But because he has become a member of a small minority, often ridiculed, he should beware of the evils of intellectual conceit and intolerance. He will not fall a prey to such arrogance if he does not forget that when "the man-in-the-street" was a child he also had his dreams, but that for him these dreams were soon lost in the dark clouds of a hostile society. Their storm whirled the dried leaves of knowledge bewilderingly about his mind so that, amid the tumult of his machines, he must say: "I must forget my dreams if I must live."

[1] From *Tristram*, by E. A. Robinson.

CHAPTER IX

WAR

"FOR here are right and wrong inverted; so many wars overrun the world, so many are the shapes of sin; the plough meets not its honour due; our lands, robbed of the tillers, lie waste, and the crooked pruning-hooks are forged into stiff swords. Here Euphrates, there Germany, awakes war; neighbour cities break the leagues that bound them and draw the sword; throughout the world rages the god of unholy strife: even as when from the barriers the chariots stream forth, round after round they speed, and the driver, tugging vainly at the reins, is borne along by the steeds, and the car heeds not the curb."

—Virgil, Georgics I.[1]

In 1939 the uneasy truce of the nineteen-twenties and thirties, that peace that knew no peace, was finally and irrevocably shattered, and for the second time in a single generation war was loosed upon the world. For six years Europe had been living in growing apprehension, or with an optimism that refused to face facts, while politicians strove vainly to prop a rotten structure tottering to its inevitable downfall. In Africa, in China, in Spain, Austria and Czechoslovakia cracks appeared which were beyond their power to heal.

Fundamentally the cause of the present struggle is the same as that which produced the last, and the fact that it is more inhuman, more barbarous and destructive, is due merely to the "inevitable progress" of western civilization in the intervening years. To the dispassionate eye of the historian of the future this war, like the last, will not appear as a clearly defined issue between the forces of right and wrong, of good and evil, which we assume it to be, but the tragic and inevitable disruption which must overtake any civilization built upon the false philosophy of materialism. For, as we have seen, a society devoted to the pursuit of material gain and the abrogation of individual responsibility is essentially unstable, and its relapse into the chaos of the law of the jungle

[1] Trans.: H. Rushton Fairclough.

can only be prevented by a form of government which, by ruthless control of individual freedom, harnesses material aims in the name of the State. The result is a predatory state, and the so-called totalitarian system of government represents no reversion to a mythical "mediæval tyranny" as many prefer to suppose, but the entirely logical outcome of scientific progressivism.

Although we claim to fight against totalitarianism in the name of a freedom symbolised by the democratic form of government, such a clear-cut distinction between modern democracy and totalitarianism exists only as an abstract principle in the minds of demagogues. In actual fact a democratic government which exists to protect the interests of predatory materialism is but one perilous remove from the predatory totalitarian state, and the one may merge almost imperceptibly into the other. We have already analysed the inevitable process by which "private" commercial interests operating in economic anarchy tend, by a continual process of coalescence, to form fewer and larger groups. Sooner or later there comes a time when these groups achieve such power and stature that they either exercise a preponderant influence on the policy of government, or compel the government, if it is to maintain its ability to govern, to exercise some form of control over its resources and productive powers. "Private" monopoly thus becomes State monopoly and, by the same token, the State assumes an increasing measure of control over the life and work of the common man. This is the unbreakable chain of circumstance by which Democracy assumes authoritarian powers and denies those very human liberties which it purports to defend.

For a nation to engage effectively in the modern and total scientific warfare of machines it must mobilize and co-ordinate its entire resources of man power and productive capacity, a fact which gave the Axis Powers their initial advantage, since the totalitarian system is the only efficient political machine for such a purpose. Faced with this threat the democracies have been compelled to marshal their forces on similar lines, a course of action which has enormously accelerated the rate of their transition from private to state control. Thus we find ourselves in the melancholy position of being compelled to perpetuate evil in order to combat evil, of battling with a spectre that knows no earthly frontiers but dwells in the camp of friend and foe alike, and whose most monstrous manifestation has sprung from seeds we have ourselves engendered. Such circumstances cause bewilderment, doubt and guilt, feelings which man seeks to resolve by attributing them to one simple overt source, some facile scapegoat, which will act as a mental sheet anchor. The disease of anti-semitism is a good example of this.

Under conditions of total war all of us, from the man in the battle line to the farmer who works to fill the soldier's belly or the doctor who heals his wounds, are concerned directly or indirectly in the business of slaughter. It is because, consciously or unconsciously, we were aware of the fact that we were faced with no clear issue of white versus black, that we took up with no more than stubborn fatalism our grim burden and faced the state-perverted ant-men of the Axis powers. Those most directly concerned in the fight know little bitterness or disillusion because, having never known the sense of security of the Victorian age, they have few illusions to lose ; nor can any amount of propaganda or political rhetoric provide them with valid principles for which to fight and so inform them with a crusader's ardour. Instead, since men in common peril instinctively evolve and cling to some form of philosophy, there has emerged among them a strange, mystical and almost incommunicable belief that to fight and die is right, being a form of expiation. In this regard, this acceptance of a probable impending death bears no relation whatever to the shallow heroics and sentimentality of perverted nationalism which have become no more than propaganda fiction. On the contrary, it approaches far more closely to the primal simplicities of Pagan and early Christian faith, when death was but the prelude to renewal. As yet it is incoherent, an emanation of emotion and imagination which cannot be reconciled by the reason with any form of modern philosophy, a shadow upon the mind's dark glass, born of the bewilderment and loneliness of the sea, the desert or the upper air. It may fade or it may grow to prove itself the true seed of renewal sown amid the ruins of Europe. From it may spring humility, and with that humility a clearer perception of absolute truth which will reveal the falsity of all our material philosophies. To do so, however, it must form coherent principles, and before it can effectively implement them they must be translated into realistic terms since to denounce the false is not enough ; its place must be taken by the true ; the best must no longer lack conviction, but acquire something of that passionate sincerity which has, for the past three hundred years, informed the worst.

All this is pure conjecture though this much is certain; that only the emergence out of this war of such a spiritual renaissance can restore to us the lost art of living and arrest the march of our unhappy civilization towards the ultimate chaos and dissolution of that Dark Age foretold by Yeats, or into the slave state of scientific technocracy. This is the real issue which is at stake in this war; it is a fight that knows no frontiers for its battleground lies within the mind. Judged by the measure of its magnitude the physical clash of warring powers, and the bombast of dictators and statesmen, are no more than sound and fury, signifying nothing, for it is no less than a grim and desperate struggle against almost overwhelming odds for the survival of the human spirit. We cannot forecast its outcome, though we can measure the strength of the opposing forces, considering first the powers of darkness in their concrete manifestation.

Pre-eminent symbol of these dark powers is the swollen bulk of scientific industrialism which, conceived by human greed instead of need, can never wax so fat or so powerful as when war provides a market for engines of slaughter and racial suicide which it cannot saturate. The impetus to its growth, which the last war provided, has already been remarked, and for the same reasons the extension of its power in the present struggle has been even greater. Physically, the war is really being fought out between these monstrous organizations, and the fighting man is but a puppet at the controls of their machines. Meanwhile war's insatiable demands upon man-power have once again brought unskilled "green" labour of both sexes to these factories in great numbers, and as a result mechanization of labour, with its attendant evils of specialization and division of responsibility, has developed to such an extent that in many large plants the technocratic "queen cell" of skilled workers has shrunk to a figure as low as three per cent. in proportion to the total labour force. At the same time in nearly all cases productive capacity is far in excess of its pre-war rate. Even if we choose to disregard the damnable influence of this apotheosis of mechanization on the individual worker, its purely material significance is inescapable. For if the pre-war economic strife and civil restlessness which led up to the present cataclysm were the result of the utter lack of relationship between industrial resources and human needs, on what a vast scale these conditions will be aggravated after the war by the world-wide extension of those resources. A further chapter of disaster can, in these circumstances, only be averted by international control of industry, by a sweeping reduction of working hours to ten or twenty hours machine-minding per week to secure full employment, and by the complete mechanization of life with cheap and short-lived goods in order that the wheels of the monstrous machine may continue to turn. These ends are by no means easy of attainment; if they fail the result will be anarchic, if they succeed the first technocratic world state of ant-men will have been born and the individual will perish.

When looked at in this sober and dispassionate fashion the prospect of the continued pursuance of the old policy of scientific progressivism after the war seems such crass and transparent folly that a change in the structure of society appears inevitable. Such an optimistic assumption, however, unfortunately fails to take into account the enormous power of the economic vested interests involved, interests which, far from being shaken by the war, are the more securely entrenched. That many are ripe for state control matters little to them, for such a change would be slight and redound ultimately to their advantage. From exerting a preponderant influence on the policy of the state they would themselves become the state by moving to Whitehall, and already at the head of the various ministerial controls their high executives are preparing to assume such office.

It is also noteworthy here to remark that, despite an acute paper shortage, their system of propaganda by press advertising has been permitted to continue almost unabated, even though they have nothing to sell to the public. The tone of this propaganda displays the anxiety of big business to pose once more as the saviour and enlightened leader of mankind, in war a romantic crusader leading "the battle for freedom", in peace a philanthropic organization concerned solely to lead us all to Utopia by the quickest and cheapest route.

Beside this acceleration of the growth and power of industrialism the other evil manifestations aggravated by war conditions tend to pale into insignificance, although they are of closely related importance. One is the wholesale breakdown of family life, accompanied by the further decay of the domestic arts, due to great movements of population under state compulsion, and the large scale employment of women in factories and services. Another is the great extension of state bureaucratic control over the life and work of the individual. This last is accepted with half-humorous resignation as one of the inevitable hardships ot war, and we are apt to forget that such control, imposed so speedily in the expediency of war, is not so readily relaxed in the peace which follows. Still less do we realize the underlying cause which has created the necessity for this vast bureaucratic machine of forms and petty officials. The key to this lies in the old adage that a people get the government they deserve. Owing to the specialization, separateness, and therefore irresponsibility of the materialistic way of life in our urban civilization, the law of the jungle prevails, and we can all commit acts detrimental to the welfare of our fellow men without being aware of their ultimate consequence. In times of peace when waste, extravagance and greed, serve to oil the wheels of industry, this total lack of moral responsibility is conveniently ignored. It is no longer possible to do so, however, when war creates conditions of scarcity and hardship amongst a people grown instinctively selfish and predatory. To limit the exploitation of the weak by the strong it then becomes necessary to enforce a measure of equality by imposing arbitrary and rigid bureaucratic control, and once again we move a step nearer to the slave-state.

So much for the forces which militate against the survival of the individual. This chapter can now be concluded in a more cheerful vein by considering those symptoms which go to prove that, though hard pressed, he is not yet defeated. Of these, perhaps the most heartening is the remarkable absence of that shallow patriotic jingoism which survived in the war of 1914-18. Neither the rhetorical political propagandist, nor the popular press with its flag-waving heroics and half-hearted attempts to breathe hate, represent the voice of the people, the majority of whom have now come to regard war as a great and, to most of them, incomprehensible tragedy, a tragedy too profound for heroics or hatred. This new and more enlightened attitude has been largely brought about by the manifest inhumanity of modern warfare, the horrors of which are no longer confined to the fighting man, but are made terribly evident to the whole people. Out of this horror is born the stubborn resolution that such a catastrophe shall not happen again, and the realization that the very fact that it has happened so soon after the first terrible lesson of 1914-18, when politicians uttered the same glib promises, indicates that there is something seriously wrong with their pre-conceived notion of civilization. As a result the average man is no longer politically minded in the narrow, nationalistic and orthodox sense of the word, but is doing more hard thinking about past, present and future, than he has done for many generations. At present he feels hopelessly bewildered by the tragedy, the chaos and the babel of ideas that assails him on every side, but the very fact that he is questioning those things which he hitherto accepted without question is vastly encouraging.

In this new and wholly admirable mood of enquiry and criticism he is helped by the fact that in certain directions the exigencies of war have necessarily called a temporary halt to the tide of "progress" which had hitherto borne him unresistingly forward. He has time to survey the course he has followed, to look with something of the historian's detached vision at the arterial roads and sprawling cities, at the jerry-built garages and road-houses already falling to ruin. As a result there is a spate of planning which must be unprecedented in history. Practically every society, religion, political party, profession or trade union has produced a plan for the post-war world so that it is with no little hesitation that the writer ventures upon an already overcrowded stage. The fact that these plans put the state before the individual is a matter to be considered later. It is sufficient here to welcome them as an earnest of sincere resolve to build a better civilization, and as a sign that the fatal Victorian philosophy of *laissez-faire* has finally spent its last breath.

Yet another source of encouragement are the growing signs of a renewed interest in the arts outside the cliques and coteries to whose ranks they have for so long been confined. For this the activities of the Council for the Encouragement of Music and the Arts have been largely responsible. Even though its activities wear the aspect of bread-and-circuses for jaded machine-minders, the fact that it has made the fine arts accessible to many for the first time cannot fail to have a beneficial influence.

Again, the evacuation of the poor of the industrial cities into rural and residential districts has forced upon the wealthier classes a knowledge of the appalling conditions which a hundred years of industrialism has bred. In this way a complacent middle class have been reluctantly compelled to consider social problems they had hitherto contrived to forget or to ignore. That one class could so long have lived in ignorance of the condition of another is itself a revelation of the separatism of modern life.

Last, but by no means least in this list of hopeful portents, is the attention which necessity has compelled us to devote to the neglected fields. It is a measure of the falsity and unreality of urban life and values that a submarine blockade and the threat of starvation was necessary to drive home the lesson of our ultimate dependence upon the country we had betrayed, robbed and strumpeted. Through the force of this necessity we are now exploiting the fertility of these neglected acres like an overworked drudge, subjecting it to the tyranny of machines and the dope of chemicals instead of nursing it with husbandmen. Here again, however, criticism should be for the present withheld since we are compelled to adopt this policy through force of circumstance. For the moment let it suffice that the town has remembered the country it has forsaken. This at all events is a step towards grace. If, after the war when man-power is no longer a problem, we can progress a little further toward the recovery of wisdom by realizing that in these fields lies our only true source of wealth, and that they offer, not a means of profit, but a way of life greater than that of the machine and the counting house, then indeed we may find a way to redemption and a lost generation will not have died wholly in vain.

PART III

There was a tower that went before a fall.
  Can't we ever, my love, speak in the same language?
Its nerves grew worse and worse as it grew tall.
  Have we no aims in common?
As children we were bickering over beads—
  Can't we ever, my love, speak in the same language?
The more there are together, togetherness recedes.
  Have we no aims in common?
Exiles all as we are in a foreign city,
  Can't we ever, my love, speak in the same language?
We cut each other's throats out of our great self-pity . . .
  Have we no aims in common?
Patriots, dreamers, die-hards, theoreticians, all,
  Can't we ever, my love, speak in the same language?
Or shall we go, still quarrelling over words, to the wall?
  Have we no aims in common?

—Louis MacNeice (1942)

Have I no harvest but a thorn
To let me bloud, and not restore
What I have lost with cordiall fruit?
    Sure there was wine
Before my sighs did drie it; there was corn
  Before my tears did drown it.
Is the yeare onely lost to me?
  Have I no bayes to crown it,
No flowers, no garlands gay? all blasted,
    All wasted?
  Not so, my heart; but there is fruit,
    And thou hast hands.

—George Herbert (1593-1632)

PART THREE

CHAPTER X

A PLEA FOR THE INDIVIDUAL

THE unique gift possessed by man is his conscious power of self-expression, and for this reason he is by nature an individualist. He is impelled to use this gift by an instinct which springs from the unconscious, or more rarely conscious, desire to transcend his mortality by moulding in the pliant but imperishable wax of the world some image great or humble, abstract or concrete, which will outlive him to record that his brief stay was not made in vain. All the great achievements of mankind and every contribution man has made towards the improvement of the common lot of humanity can be attributed to the proper fulfilment of this function. Yet because the gift of self-expression which makes these achievements possible, necessarily involves freedom of will, the creative instinct which prompts its exercise is subject, like the sexual instinct, to perversion in the shape of a deviation of aim and object. The deeper spiritual significance of this will be considered in a later chapter, it being sufficient here to remark that the analogy between creative and sexual instincts is, in fact, very close, since both are concerned with regeneration and are therefore manifestations of the eternal association of immanence with transcendence, of decay with renewal. Similarly, the fruit of their perversion is evil because it is mortal, sterile, and therefore contrary to unalterable natural law. In each individual the normal instinct is initially dominant but, because of the freedom of will, the tendency to perversion is also present in latent form. The reason why perversion can so frequently grow to dominate the normal is not yet fully understood. We can only say with certainty that perversion must obviously occur through frustration if education, opportunity or environment conspire to hinder or deny normal expression. It therefore follows from this that while we cannot eliminate the tendency to evil in the individual without eliminating free-will, a course as undesirable as it is impossible of complete attainment, we can foster and encourage the initial tendency in favour of the good by setting ourselves to foster conditions in which the normal creative instinct can be most freely exercised.

The perversion of the creative instinct is the "will-to-power", a substitution of the normal aim of spiritual and immortal fulfilment by the perverted one of a material and mortal desire to dominate. This material ambition is the root source of all evil just as the normal instinct is the source of all good, and thus both good and evil spring from the same root to afford yet another instance of that eternal association of opposites by which all things are measured. The recognition of this is implicit in the teaching of the greatest philosophers, and particularly in that of Jesus where it is surely the key to those statements of apparent paradox which have been the subject of so much theological controversy. It is also evident in crude form in the symbolism of so-called pagan religion where the conception of god incarnate in man represents man's unique and immortal potentiality, while the ordained death of the man-god after a prescribed cycle and the bestowal of his divine attributes to a successor is a symbolic reminder of the transience and vanity of the mortal and material.

Throughout the middle ages human conduct was controlled by the tacit acceptance of these Christian-pagan tenets. Though the period was in many ways evil and violent such manifestations were *overt,* their nature was realized and the transgressor could seldom escape the consequence of his action and the recognition of his guilt. As the religious view changed, however, these principles became obscured and consequently the distinction between right and wrong clouded. We have already observed how the power which the church wielded brought about its own perversion and corruption, and, by attempting to restrict the freedom of the intellect, itself a normal outlet of the creative instinct, encompassed its own downfall. We have also seen how the new doctrine of the will-to-power developed after the Renaissance until the liberty to exercise this perverted aim was confused with freedom and, under the title of "enlightened self-interest", became the philosophy which was the driving force of the industrial revolution.

It should now be evident to all but the blindest reactionaries that this philosophy is false, and that the social structure it has built up in the name of this vaunted freedom has produced conditions of slavery for the majority and is also as unstable as it is complex. In this complexity the individual has been submerged and has lost his responsibility with his freedom, with the result that he not only escapes, but is ignorant of, the consequences of his conduct. Furthermore, since his creative instinct is denied the freedom of ability in which alone it can function, it becomes perverted into "ambition" as we prefer to term the will-to-power. In these circumstances any clear conception of right and wrong becomes impossible and, consequently, although the symptoms of the disease of our civilization have become painfully apparent, we are unable to distinguish cause from effect or to isolate the microbe. We are reluctant to accept a diagnosis which might involve drastic surgery, preferring to keep the present body intact and to cover its sores with the palliatives of collective humanism.

The many plans for the post-war world which are being made to-day in the sincere desire to remedy past defects have one common denominator in that they consider the community primarily and the individual secondly. They are thus merely an inversion of the philosophy of the Manchester school since they declare that the self-interest of the State will automatically secure the well-being of the individual. Unfortunately, however, because the whole complicated machine of modern civilization is the product of the materialistic perversion of ambition, it can only function upon that principle, and therefore the humanist's efforts to secure justice within this framework must necessarily be confined to a material end, to securing freedom from want and from fear through economic security. Furthermore, this negative freedom can only be achieved by enforcing such measures of control over the individual in the name of the State that not only is his ambitious perversion curbed, but he is also deprived of a normal outlet for his creative instinct. He is thus denied the only real and positive form of freedom, the freedom of ability without which the human spirit withers and dies. The conception of freedom promulgated in the much vaunted "Atlantic Charter", and in defence of which we are said to be fighting, is in fact purely negative in value as can be demonstrated by simple analogy. For it would be possible to condemn a man to solitary confinement in a locked room for life and yet, by satisfying his every material need and giving him every possible security from violence, accident and disease, to conform by such treatment with every article in the Charter. There can be little doubt that a man so confined would be unlikely to appreciate the consideration of his gaolers, his lot being precisely similar to that of the unfortunate Mrs. Frankford in Heywood's comedy, *A Woman Kilde with Kindnesse*. If his spirit was not broken he would resist with increasing violence the ministrations of his gaolers, nor cease in his efforts to escape, thereby laying himself open to those very perils from which they were concerned to protect him. This, in brief, would be the lot of the individual in the planned world envisaged by humanists who accept the present social structure. If the human spirit perishes their plan may be achieved, if it survives the result will be further violence, anarchy and social disintegration.

What then is the answer to the problem? Obviously there is no easy way out. That is the first thing to grasp, for it is a measure of the extent to which we are all committed to the fallacy of materialism that we are forever pursuing quick results of little worth and ephemeral value. We think of ourselves and not of our children or our children's children ; we plant Sitka Spruce where our forefathers planted oaks. No plan of our mortal devising can undo in our time the accumulated evil of centuries, but we can sow the seed of a new and better civilization to-morrow if we wish, though it may be two hundred years before it come to flower. What greater end could we pursue? What more could we desire of life than that in some far distant time man should look backward down the years to bless and say of us that in our darkness, undismayed by the wreckage of our civilization, we set about the building of their own?

As a first step it is of absolutely vital importance that we learn that all planning for the future must start with the individual and not with the community. The whole exists in the part, and any plan which ignores this truth can possess no foundation in truth. A community cannot be stable unless the individuals of which it is composed are leading full and contented lives, and no material amenities or social justice can compensate for lost content. Surely, therefore, the aim of individual contentment is one upon which all can agree, however they may disagree over the means by which it may be achieved. Such disagreement is in great measure due to the fact that our civilization has lost the key to the true meaning of the word content, and is inclined to confuse it with the vicarious pleasure to be derived from the enjoyment of material success or mass-produced amusement. True content is a quality of the spirit which cannot be analysed and defined, a quality which is the issue of a harmonious marriage of knowledge and wisdom within the human mind and therefore the reward of the good life. We can only seek to recover this lost quality by a close study of the spiritual as well as the material needs of the individual before attempting to create an environment in which these needs can best be satisfied.

Man's material needs we can enumerate without undue difficulty. He has a right to equality of opportunity, to a just portion of the earth's bounty, to freedom from oppression and violence, and to security in his old-age. His prime spiritual need, however, is the freedom of ability which will enable him to satisfy his creative instinct for the enrichment of his own life and that of his fellows. It is this freedom which we must set ourselves to create, for the more widely it is achieved the less becomes the risk of that perversion of material ambition and "will-to-power" which is the source of all evil, while without it there can be no true contentment.

In attempting such a task it must ever be borne in mind that the creative instinct finds satisfaction primarily in the actual work of creation, and only secondarily in the result achieved. The dream of the mechanistic "Leisure State" is Utopian and fallacious simply because the modernist is blind to this simple truth. If man's environment enables him fully and freely to exercise his creative function he will, owing to what W. B. Yeats called "the fascination of what's difficult", invariably set himself a standard of perfection that is always a little beyond the range of his ability to attain. This does not mean that the goal of creative work is barren. Far from it. It not only contributes to the real wealth of the community and to the contentment of the creator, but at the same time it acts as a stimulus to further and higher effort. This pursuit of quality for its own sake is the noblest manifestation of man's unique endowment, and its continued development represents the only true form of human progress. When and wheresoever the arts of man have in the past developed to a high degree of perfection, such perfection has been achieved in one way and in one way only, by a hereditary succession of craftsmen each advancing the work of his predecessor a little further towards an unattainable goal. There is no alternative to this progress of qualitative development, and any form of "progress" founded upon any other aim than the qualitative is not progress at all, but retrogression, which inevitably leads to the decay and disintegration of society. For so soon as the creative instinct becomes perverted into the will-to-power and, therefore, becomes harnessed to the pursuit of a material end, not only are the social consequences disastrous, but the results achieved by the individual are as utterly barren as the means are uncreative. Therefore the frustration so caused can only be assuaged by the automatic pursuit of yet more power. The craftsman's qualitative ideal which results in the creation of real wealth, contentment and the good life is thus transmuted into that "vaulting ambition that o'erleaps itself" in the quest of elusive fruit which becomes gall and wormwood so soon as it is plucked.

Firstly then, with this qualitative principle in mind, we have to consider how, by education, we can best foster the normal development of the creative instinct in the individual, and secondly, how we can so modify the structure of society that the individual will be able to take his place within its framework without sacrificing the freedom to exercise and, still further, to develop that creative ability which he has thus acquired. Only by fulfilling these conditions can we hope to steer a safe course between the Scylla of the Servile State and the Charybdis of chaos.

Chapter XI

EDUCATION FOR FREEDOM

To-day there exists a considerable weight of opinion in favour of equality of opportunity for education, although there is no clear conception of the form which that education should take. We have begun to realize that the best facilities for education which a nation can offer should not be the exclusive privilege of wealth, but should be available to all; that the so-called Public School should justify its title. At the same time it is becoming apparent that there is something seriously wrong with pre-conceived notions of the form of education and that, all too often, so far from encouraging individual ability, education has frequently been a stultifying influence from which, if it has not been irrevocably damaged, ability has painfully emerged in later life. Yet, despite its gross inequality and its haphazard method, the faults of our existing educational system have in the main been those of omission, that is to say it has not deliberately restricted individual development, but merely given it no help or unwittingly confused it. Its cardinal sin has been its tendency to inculcate the idea of ambition as a virtue instead of the perversion which it is. In the light of such an idea the impressionable adolescent regards acquired knowledge, not as so many useful tools which he must learn to use to perfect his life and work, but as weapons with which to arm himself against his fellows in order to secure for himself a good share of the spoils in "the battle of life". By such means is the evil of the will-to-power not only perpetuated but encouraged.

It is obvious that this careerist incentive is incompatible with the planned society of the humanists in which the interests of the individual will be subordinate to those of the community. Yet in a technocratic state where uncongenial, mechanical and mindless tasks await the majority, the economic incentive can only be replaced by a form of compulsory service. The automatism of machine shop or assembly line remains the same whether it is controlled by "private enterprise" or by the State, and if by enlightened education the potential ability of the individual is encouraged he will not then voluntarily submit to such subordination. This is a problem to which the only answer is that such an education cannot be reconciled with the present materialistic framework of society, and will therefore probably be withheld for the same reason that has prompted the present ruling-class to withhold it.

We are therefore faced with the terrible danger that the present clamour for education will lead to no renascence of the human spirit, but will take us another fatal step along the broad and all too easy path that divides a nation of predatory individuals from a predatory state. For, in our present society, the only effective substitute for the economic incentive of the will-to-power is the idea of work as a service to the state, a doctrine which implies the deification of the state, the renunciation of all individuality and complete subjection to the will of the technocratic "queen-cell". Those who doubt that such a pernicious ideology can ever be inculcated on a national scale should consider conditions in the Fascist countries and in Russia before returning their dusty answer. In fact, the masochism of this self-abasement is but the inversion of the sadism of the will-to-power, and both are thus twin aspects of that perversion which results from the denial of the freedom of self-expression.

Thus in our attempt to undo the result of three centuries of self-interest we stand in deadly peril of creating an even more terrible manifestation of evil, the Frankenstein monster of the Machiavellian machine-state built upon the deliberate perversion of the human spirit. Never in the whole course of recorded history was that spirit in such dire peril as it is to-day, yet to-day we can still save it, while to-morrow may be too late. What could be more fitting than that we who led the world into perdition by pioneering the industrial revolution should now set the course toward redemption? We can only do so if all who value the freedom of self-expression as man's supreme and inalienable right abandon their sectarian differences, their fatalism and apathy, and combine in stubborn defence of this freedom. They may not be able to enter the muddy political arena, the past has shown that they are no match for professional demagogues, but they can put on the armour of their particular ability to defend the coming generation by ensuring that as many as possible shall receive such an education that they will not lightly submit to state tyranny.

For this task the artist is best fitted either by direct influence or through the medium of his work and that of his forerunners. This may seem a humble, prosaic, and unromantic assignment, yet it is, as we have seen, a vital first step upon which the future of the human race for a thousand years may depend.

We have already established that the fundamental principle of enlightened education must be to encourage and develop the creative instinct in the individual. This is a task requiring sympathy and understanding, a difficult task, but a straightforward one. In early years the child should not acquire a distaste for education through having disembodied facts, dead languages or the dates of forgotten battles drilled into his head on the outworn plea of "mental discipline." No knowledge should be imparted to the child which cannot be associated with some aspect of the child's life, environment, or practical activity, nor should the teacher ever fail to make clear this relationship. In this way and in this way only can the child's interest be ensured until such time as he begins to reveal that discrimination which may indicate to the discerning teacher the emergence of particular ability. There can be no foretelling when this "bent" will first appear, it may do so at an early age, it may not emerge until the child is almost adult, or there may be several false starts, for the child should never be compelled to pursue a course which he has discovered to be uncongenial. It is when this stage is reached that the really vital part of education should begin. For the average child of working class family, education ends in early adolescence, generally before ability has begun to emerge, and the best course then open is a period of apprenticeship in industry, accompanied perhaps by "night school" at a technical college. Unless the parent can afford to pay a substantial premium the modern employer all too often abuses the principle of apprenticeship by regarding the apprentice merely as a cheap labourer or potential machine minder, while after the day's work the apprentice is too weary at evening classes to pay serious attention to theoretical teaching which he is unable to associate directly with his work. It is of paramount importance, however, that the closest possible connection should be maintained between theory and practice. Therefore, while nothing could be better than apprenticeship under a wise master, it would be preferable, so long as the present industrial conditions obtain, to provide practical training in school workshops rather than risk the severance of this vital link. For it is only by preserving this close association that the individual can become the skilled master of his vocation.

Even if this task be accomplished, and even if the adolescent is taught to regard perfection of his work as an end in itself, education will merely have placed in his hands the tools of knowledge without indicating how best those tools should be used to contribute towards the good life of the individual and of the community. This right use of knowledge is wisdom and without its leavening the greatest acquired skill and knowledge are but the " dry leaves under the tree ", and may be perverted to base use. On the contrary the simplest task, if it is informed by wisdom, can contribute to the good life. But wisdom, as has already been stated, cannot be taught; yet, if we bring wisdom to the task of education, we can create the conditions most favourable to its growth, and it is into these conditions that we must now enquire.

They can be summed up in the one word—association. Just as in the early juvenile education nothing should be taught to the child without being related to the child's environment or experience, so in adolescence all imparted knowledge should be related to the individual's chosen vocation and environment. At first reading this may appear to be an argument in favour of specialization; in fact it is the reverse, as will become apparent if we bear in mind the truth that the whole exists in the part. The individual can never derive proper satisfaction, nor acquire true content from life if he cannot associate the exercise of his particular ability with the other aspects of his life and that of his fellows. His work, his leisure, his home life and his environment, must be related with each other and with the same facets in the life of the community until they appear inseparable parts of one whole. This is association in space which, in turn, must be intimately linked with an awareness of a corresponding association in time, that is to say the individual must be intimately conscious of the work of the past that has contributed to all his associations in the present, in other words he must acquire the historical sense which is the basis of all tradition. Only by such means can the individual consciously or sub-consciously become cognizant of his place in space and time, and by so doing establish within his mind that relationship between the relative and the absolute, the temporal and the eternal, which is wisdom.

If by analogy we compare these associations with so many planets revolving about the human mind in settled orbit, as about their parent sun, we then see that the spread of knowledge in the last three centuries has, so to speak, weakened the gravitational power of the mind with the result that there has been a general and rapidly accelerating disintegration of an ordered system. This disruption has taken place in an outward direction, the planets, released from their common orbit flying apart and splitting up into bewildering galaxies of disorder, no longer obedient to their central sun, but moving in independent and unpredictable courses. At times one or more may swim into its orbit, but the ordered system cannot be restored. When man's field of associations has disintegrated in this way, as it has done in our social order, it is useless to prate of whole nations as though they were intelligent or responsible groups, of wider horizons and possible world states, simply because our scientific inventions can now encompass the material world. The idea of the nation or of the world can only be made real for the individual through the progressive widening of his existing and more intimate orbits of association; if these associations do not exist there can be no such extension, and these "wider horizons" remain the fond illusion of political propagandists. In our separatist society nationalism and internationalism are abstractions no matter how earnestly we may try to convince ourselves to the contrary. Once again the whole exists in the part, that is to say the nation for the individual is represented by that part of it with which he is most familiar. Thus England, if it possesses any reality for the Englishman, is represented primarily by his home, and secondarily by the region in which that home is situated. These domestic and regional associations include not only the natural characteristics but the way of life, the speech, customs and traditions of the people within the region which, generated and conditioned by the peculiarities of that region, constitute its *genus loci*. The strength and reality of these associations is proved by the fact that all through history it is to them that principalities and powers have appealed whenever they desire to promote patriotism and racial hatred for war. So much so that in this our day, when the way of life of which they were once the flower has almost been destroyed, these associations are still invoked as symbols to promote militant nationalism. While "big-business" plans to carry the ugly tale of the Highland evictions a stage further by flooding the Glens for power, the skirl of the pipes sounds in Africa and Sicily, and the B.B.C. features news and concerts in Gaelic. Consequently all that is meant by regionalism has become inextricably associated in the modern mind with aggressive and militant nationalism, so that the progressive humanist regards them at best as picturesque archaisms which cannot be reconciled with the cause of world peace and "the brotherhood of man". Thus the modernist is prepared to sacrifice them in the pursuit of his Utopian goal. But because these regional associations have been repeatedly invoked and misused as a potent weapon of propaganda, it is not to say that the way of life which they reflect is incompatible with world peace, understanding and social stability. Actually the reverse is true, namely, that the stability of the home depends upon the strength of these associations, and there cannot therefore be national or international stability without them. To destroy them as we are doing, to substitute for them the abstract uniformities of the machine state, and then to attempt to build upon such a foundation a world-order is to build a house of straw. Bereft of these associations which link past, present and future, and which alone can give life meaning and purpose, modern man is borne along like a ship without rudder or compass upon a flood he has created but is powerless to control, and which bears him he knows not whither, a prey to every wind of evil in the guise of dictator or demagogue, with their racial mystiques and ideological abstractions.

Because the associations of the present cannot be restored except slowly and painfully by a process of regeneration and change within the social structure, it is the temporal associations of the past that we must first endeavour to re-establish by education. Yet here we tread perilous ground for we have to dispel a cloud of misconception and false sentiment. On the one hand we are confronted by the wilful arrogance of scientific materialism which ignores the past, and on the other by the sentimental nostalgia which accepts the past uncritically. Of these the latter is the more insidious. For objective reality can readily be invested with false associations when seen with synoptic vision down the long corridor of the past. For example, it is easy to-day to sentimentalize over the Victorian era, although it is doubtful whether any who do so could produce a valid argument for its restoration. We learn from its errors rather than from its achievements. The more disruptive and separatist the present, however, the more will men turn with nostalgia to the past, the tendency being one of natural reaction. But because the link between past and present has been broken this perspective is generally distorted with the result that the work of the past, its people and events become invested with entirely false associations which are woven by romantic wish-fulfilment. This generally results in attempts to impose the concrete manifestations of that past upon a present which entirely lacks the principles responsible for their production. The Gothic Revival and the later Olde Worlde mania, which followed the Art and Craft movement, have left sufficient hollow monuments behind them to testify to the ghastly failure that must always attend upon any such attempt. The reaction to such falsity has been the tendency on the part of "progressivists" to treat any suggestion that we should not scorn the past, but attempt to learn from it, and so establish continuity and tradition, as a contemptible and defeatist endeavour to "set the clock back".

These are the two opposing misconceptions between which our education must steer its perilous middle course of sanity. The history we teach must rid itself of that unfortunate emphasis upon battles, sudden death, the rise and fall of princes and powers, and the duplicities of international diplomacy, for these are but boils upon the face of time caused by the poison of the will-to-power, and as such they can have no association with the absolute and the eternal. The face itself is the history of the common man through the ages, his hopes and fears, his work and leisure; it is a history illustrated, for those who have eyes to see, by countless memorials, the deathless, eloquent marks of their labour which we now neglect, destroy or defile. It is of these our forebears that we must speak, telling of their wise village organization of common field, court-leet and manorial custom, of how they ploughed, sowed and reaped and celebrated that fruitful partnership between labour and natural bounty in gracious traditional ceremony, whose symbolism was a recurring reminder of the association between decay and renewal, mortal and immortal. We must speak of how they utilized local resources and materials in true economy, of their tools and all articles of use and how cunningly and lastingly they wrought with them to build cottage, barn and mill, wagon and plough, as "men upon some Godlike business". In this our task will be made the easier for the efforts of that enlightened few who, in an age of iconoclasm unparalleled in history, have laboured to preserve and record these things that they might not perish utterly. We must emphasize that this rural and communal organization was no crude archaism which decayed by a logical process of enlightened supercession, but a delicate system evolved through many centuries of slow and logical adjustment to human needs and capable of a far finer flowering, which was finally destroyed from without. Admirably fitted for its purpose like all its products it rode the storms of centuries more lightly than any other of man's experiments in social organization have ever done, conquering its conquerors, staunchly defending the inalienable right and freedom of the common man to his share of the common earth until, after a thousand years and in the darkest hour of our history, that freedom was finally wrenched away. This tragedy is not to be found in our history books, at the best it is condensed into a short dry chapter called "Enclosures". Because the evil which provoked this crime still holds the reins of power, our civilization does not deem it wise to tell its children the true story of how their forefathers lost their freedom with their fields. How, bewildered, exploited and starved by dark forces they could not understand, they rose to defend their own with stubbornness, but moderation, born of age-old courtesy, and were broken by the ruthless savagery of exile and the rope. How those who escaped this fate remained as slaves on the land they had lost, or drifted to the abject poverty and squalor of swollen cities where, until the early years of this century, they were not even entitled to the ignominy of Poor Relief unless the family earnings were less than one shilling and ninepence per head per week, after payment of rent. This did not happen in some remote "dark age", but within the memory of our grandfathers. Our history books do relate how at this period we made great strides in science, engineering and commerce, and were building up our empire, but it is of this deadly canker at its root that we should speak.

This teaching of the history of the life of the common man must be intimately related to the particular district to which those whom we are teaching belong; to its geography, geology, botany and natural resources, with which, because it was an organic life, it is inseparably associated. Kings and violent men with their marching armies have vanished, leaving little trace other than breached walls and half-remembered legends, but there is not a town or village, not a field, a barn, a hedge or a tree, that cannot speak to the children of their ancestry, provided we give them the ears to hear and the eyes to see.

This method of education by local association and historical approach would not create parochialism in the narrow modern and derogatory sense of the word because, although the concrete manifestations of past life vary in fascinating sequence from village to village and region to region, the principles which informed that life remain the same; that is to say the variations represent the logical adaptation of universal principles to meet local conditions and local needs. For this reason the history of the English village through the ages is the history of the English people; more than that, it is the history of the peasantry of Europe, it is the whole existing in the part in accordance with natural law and truth.

Supposing for a moment that we were able to institute a national educational system upon these lines, who then, it may be argued, would be left to do the "dirty work"? If by dirty work we mean the soulless mechanical tasks of mechanized industry the answer is no one, and if we mean the hundred and one parasitical and uncreative occupations which, in our separatist society, fill the gulf between producer and consumer, the answer would again be the same. But if by "dirty work" we mean the crafts of the manual labourer, the answer is that there would be many. For it is these tasks which have the longest and proudest history and it is to them that we should restore the dignity we have contemptuously destroyed.

These principles and aims for the education of future generations apply to women no less than to men, and they should be accorded the same freedom to develop and to exercise their particular ability in any field upon terms of complete equality with the opposite sex. Yet, because of the method of education by historical association, emphasis in the woman's case would naturally fall upon those domestic arts which share with the work of the fields the proud distinction of the oldest lineage. As we have seen in an earlier chapter, the hearth was once not only the woman's workshop, but was so intimately linked with all the activities of man's life and work that it was the nucleus and therefore the central symbol of all the associations of that life. It was because the hearth gave way to the falsity of the "front room" that woman forfeited her spiritual home and consequently sought an outlet in other ways by clamouring for "Emancipation". Therefore, though freedom of opportunity must remain, the more effectually we restore the craft and associations of the hearth the less will be the tendency of women to pursue vocations which are the more natural functions of men.

Finally, we are faced with the problem of teaching the teachers, for all too often the task of education has been left at the mercy of time-servers attracted to the profession by the sheltered life, long holidays and eventual pension. Too many enter the profession straight from the shelter of university or training college and therefore have none of the practical experience of life, without which they cannot acquire that wisdom which must go hand-in-hand with knowledge. Again, those who have wisdom possess no skill in the difficult art of teaching children in a manner which will hold their interest. The artist, for instance, is not necessarily suitably equipped for the task since his art is statement, not argument or explanation, and so, while it speaks eloquently to those who share his conception of reality, it will be meaningless to the majority. Like the life of the past it must be explained and related before it can become intelligible.

Because so much of our hope for the future rests with him the artist must, however, attempt to overcome this difficulty, not by changing his art, for that is impossible, but by formulating and trying to express in the simplest terms the principles of which his art is the product. It is not suggested that he should sacrifice his art to the whole-time business of teaching. We should abandon the preconceived notion of the school as an exclusive seminary with an inflexible curriculum imposed by a permanent staff, and regard it instead as a storehouse and pool of local knowledge to which all should be invited to contribute their experience for the benefit of coming generations. Such a system under the wise guidance of permanent teachers capable of co-ordinating and explaining the work of others would not only open unsuspected mines of local knowledge and wisdom, but would continually contribute to the experience of the teachers as well as to that of the children. The work would be further assisted if the school were also to house a museum, where all the "bygones" used by past generations in the region could find a place and so help to link the past with the future of new generations. Here, too, modern invention in the shape of the cinematograph could find its proper function as an invaluable assistant in education by demonstrating in visual form subjects which, for one reason or another, were beyond the reach of tangible contact. It were better, however, that wherever possible the film should serve as illustration for a lecturer, and not rely upon a mechanical voice of its own which, no matter how able, cannot adapt its language to a particular audience. Furthermore, the teacher should be in control of the film so that he could at will make use of slow-motion, repetition, or the "still" to illustrate his words or to clarify some point upon which he perceives his audience is not clear. Used in this way the film could help the teacher to make real for the child the life of the people in other lands, and how, in the way they adapted themselves to local conditions, they demonstrate the same common principle.

Much has been made of the potential value of broadcasting in education, and here again scientific knowledge has given us a useful tool, but also an extremely dangerous one, if it is not used with wisdom. As a means of instilling musical appreciation when direct performance is impossible, it has great value. By its aid also the best living speakers and actors can bring the classics of poetry, literature and drama into the school in oral form. But as a method of direct instruction it would be as useless for our purpose as it would be dangerous, because it is subject to the same failing as the mechanical voice of the film. It represents knowledge, arbitrarily imposed from without, and is therefore opposed to the whole principle of the education which has been outlined in this chapter.

Assuming that we were to succeed in establishing such a principle of education, and that by its aid we were able slowly and painfully to restore the associations and spiritual values which we have lost, and with them the conception of true freedom, what then follows? How would a generation so assured of its place in time and space confront the hostile present of our existing social structure? It would assuredly set about the task of re-building that structure to conform with its ideal of the good life by restoring the broken links of continuity between the past and the future.

Chapter XII

WORK AND WEALTH

In the profound symbolic legend of the Garden and the Fall we are first presented with the picture of primal man, unaware of his unique but perilous gift, and drawing sustenance from natural resources to the existence and growth of which he had as yet made no contribution. Neither good nor evil could exist in this world within the measure of our understanding because their seeds had not yet germinated in his mind. The creative instinct, however, led him to tamper with the natural forces upon which he depended for his existence, and by so doing he became conscious for the first time of his unique ability, not only to create but to destroy. Hitherto he had been content to accept the earth and its riches as he found them, but now he could no longer do so since he realized that he could combine his conscious ability with these resources to produce a more fruitful result. By doing so he increased the yield of the necessities of human life to an extent which enabled his seed to increase and multiply to a density far in excess of that which unaided resources could support. In this way the livelihood of mankind became dependent upon the combination of natural resources plus human labour. Of these two the first is constant, pre-ordained and unalterable, whereas the latter is variable so that in fact the welfare of the human race depended, as it must always depend, except in the unlikely event of a return to primal simplicity, upon the degree of wisdom with which man employed his creative gift. This human factor is variable both as to method employed and to end desired, otherwise it would be powerless to influence or to modify the sequence of natural events, in which case it would not exist, and we could draw no distinction between man and the lower animals. Where a variation becomes possible it necessarily follows that extremes must exist as opposite poles of which the variation is the field. These poles are good and evil, and man discovered their existence when he found that he could not only make the desert to blossom as the rose, but lay waste the garden and make it barren as the desert. Thus he became conscious of the potential power he exercised not only over his own garden but over that of his neighbour.

So soon as mankind "partook of the Tree of Knowledge" and became dependent upon this combination of natural resources, plus human labour, the factor of human inequality at once appeared. This inequality was due not only to variations of human ability but also to the unequal distribution of natural resources over the earth's surface as a result of climatic and geological variation. This inequality aggravated those sins inseparable from the sense of power, greed and covetousness. In the more fertile portions of the earth such as the Mediterranean basin the development of an ordered and stable civilization dependent upon settled agriculture was very rapid, while in less favoured regions man remained a hunter, or at best domesticated the animals upon which his livelihood depended to found a society pastoral or nomadic. At the points of interaction between these two forms of society, trade by barter was established but, owing to acquisitiveness on the one hand and envy on the other, the agrarian civilizations were subject to successive waves of invasion by the nomadic peoples whose mobile and more predatory mode of life made them more than a match for the sheltered and peaceable agriculturists. Nevertheless archæological research has proved that where an agrarian civilization was, by historical accident or by natural barriers, protected from destruction *from without* it achieved, by a process of organic growth, a stability accompanied by such a flowering of the arts of life as to seem to us, in our age of transience and of ephemeral values, almost inconceivable. Of such a civilization that of the Cretans is a notable example. Researches at their capital at Cnossos have shown that this was not so much a town as a vast palace inhabited by the king and his people, and that it was equipped with water-pipes, bathrooms and, in fact, with all those material amenities which we arrogantly assume to be the exclusive attributes of our civilization. We have found also that Cretan craftsmen were skilled in metal work and in-lay, in sculpture, painting, pottery and textiles, in jewel work of ivory and precious stones. Their system of writing we have been unable to decipher so that we cannot measure their achievement in literature, drama and philosophy, though it is logical to assume a correspondingly high development since the products of applied art of a people are but the concrete manifestation of their thought.

The knowledge of these discoveries in Crete occasion surprise to many of us who, despite all the historical evidence of this and the subsequent civilizations of Greece and Rome, cannot rid our minds of the fantasy that the past represents a "dark age", necessarily more primitive than our own, from which our present civilization has developed like a pristine flower out of the dark soil of ignorance and superstition. For this illusion the fallacious assumption of automatic progress is responsible. It leads us to suppose that because the materialistic aim of our civilization has induced the unique development of a number of dangerous mechanical toys which give us hitherto unprecedented power to exploit or to destroy, we therefore stand upon the topmost pinnacle of human achievement on this planet. It never occurs to us to reflect that the reason why past experiments in civilization afford no evidence of similar development may not be due to any inherent and "primitive" incapacity, but that they pursued different aims, the validity of which we have no authority brusquely to dismiss.

Thus the evidence of the Cretan civilization, which occasions us so much surprise, is simply the logical flower of the growth of an organic society which, it is estimated, enjoyed a life of uninterrupted peace and prosperity for at least a thousand years. Cretan art was thus the result of this enduring stability which was its real achievement, and if our material achievements were the complement of a similar stability and ordered development, our pride in them would be amply justified.

When eventually the Cretan civilization fell, its downfall was due to precisely those reasons which have already been indicated. For centuries Crete traded justly and equably with Egypt; it sowed the seeds of its destruction so soon as it developed a predatory, violent and unjust trade with the Greeks, a nomadic people from the north who established themselves in the Mediterranean basin. The conquest of Crete and the destruction of Cnossos by the Greeks was the eventual and inevitable consequence.

When we look back down the long corridor of man's history, short-sightedness and the foreshortening of our perspective of its further distance creates the illusion that the remote past was one of continual war and upheaval out of which civilizations sprouted and decayed as swiftly as mushrooms. On the contrary, we have accumulated abundant evidence which proves that the Cretan civilization was by no means unique, and that, in fact, the converse is actually true, man's attempts to create an ordered and stable society becoming progressively more short-lived, ephemeral, and therefore less successful as time advances toward the present. None but the most wilfully myopic "progressivist" can fail to find this fact significant and to ponder the underlying causes which have brought it about.

As we have seen, the only real source of human wealth exists in the natural resources of the earth which represent a pre-ordained source of "capital", which is, or should be, available to all men for their use. Man's income from this capital was, however, very meagre until he discovered that, thanks to his unique creative gift, he could, by the expenditure of labour, increase it. As man perfected this ability his wealth increased, and he was able, upon the basis of this fruitful partnership between man and earth, to establish a stable form of society which could only be overthrown by the perverted notion of that organic wealth, not as an end in itself, but as a means to acquire power. But as wealth increased, so also did inequality of condition widen not only between whole peoples living in more or less favoured regions, but between individual members of the same community. The former inequality was responsible for the downfall of the early civilizations, while the latter did not prove substantially disruptive so long as the common goal remained the acquisition of real wealth. At the same time, however, trade by exchange and barter increased as man's activity became more skilled and diverse until its volume and complexity demanded simplification. The problem was solved by evolving an artificial standard of value by means of which real wealth could be measured. This took the form of a token, readily portable, and generally made of metal, then rare in any form, the value of which was equated to that of some standard unit of real wealth such as a beast or a fixed measure of grain. At first these tokens consisted of ingots of iron, silver or gold, which still possessed real, as well as representational value, in that they could be wrought into articles of use. Soon, however, they became small discs stamped with the insignia of the issuing authority, which were still more readily portable and storable, but which had a representational value only. In this way man invented the monetary system. So long as the standard of real wealth was not obscured, and coinage was only issued and expended upon this basis it could not fail to represent accurately the real wealth of the holder or the just price in an exchange between producer and consumer. Yet whereas before the introduction of this currency there was naturally a limit to the amount of real wealth which any individual could acquire because it consisted only of consumable goods or articles of use or beauty, the able or more favoured producers found themselves accumulating a store of monetary tokens the potential extent of which was limited only by their acquisitive ability. In this way man realized that in this new system he not only had a convenient representational substitute for real wealth, but he had power in a form readily assimilable. For although his real wealth, that is his capacity to produce, was no greater than before, he could now exercise power over his less favoured fellows by commanding their productive capacity in his service in return for a token payment with which they could obtain the needs of life. But although it is possible to determine an accurate and just token equivalent for the products of human labour, it is manifestly impossible justly to assess the value of that labour except in the immutable currency of its result in real wealth. But token payment for work done did not necessarily bear any direct association with real value and in consequence, man the worker, became entirely dependent for his enjoyment of real wealth, a portion of which should have been his by inalienable right, upon the human justice and morality of man the master. Naturally, the more men the master could employ the greater the real wealth produced, and therefore the greater the token wealth in the master's hands, but if the tokens expended upon the labour employed could be reduced, the master could then accumulate such wealth out of all proportion to that due to him in representation of his own share in the creative production of real wealth. In this way human labour became a liability.

Man soon discovered other methods of increasing his monetary wealth, and therefore his power over his fellow men, out of proportion to the share of real wealth which was his due. He could lend his money to his fellow merchant in return for a small but regular repayment in the form of "interest" which totalled far more than the extent of his loan, the margin representing no increase of real wealth, but a share of the token profits derived from that enslavement of others which his loan had made possible. In this way the connection between real and token wealth became much looser and more complex until man found that he could with great advantage act as an intermediary in the process of barter between producer and consumer by selling for more tokens than he bought. By so doing he ceased to make any contribution to the production of real wealth, but became a human parasite, accumulating the power of money at the expense of both producer and consumer by acquiring real wealth for less than its true value and selling it for more than that value. These were the sins of "usury", "engrossing" and "regrating" which mediæval law condemned.

Such economic sins had the effect of aggravating the condition of inequality which had up till that time only been attributable to variations of human creative ability or variations of climate and natural resources. As a result the task of developing a stable form of society became increasingly difficult. Nevertheless for a long period following the introduction of currency the true conception of wealth as a product of the organic association of man and earth remained dominant, for man had not yet solved the physical problems which were to make possible the large scale exploitation of both for the sole sake of acquiring money power.

Thus it was not until the rise of the Roman Empire that a social system emerged which bears a close resemblance to that with which we are familiar. Here for the first time in history the associative link between man and the earth which had served to remind him of his eternal dependence, was broken on a large scale. The heart of the Roman Empire was unique in that it was urban, in the modern sense of the word. It consisted of a large and dense population concentrated in cities having no direct association with the sources from which it derived its real wealth and therefore concerned only with money which, because of this dissociation, acquired its own fictitious value. This urban civilization developed to an extent far beyond the capacity of local resources to support. It was therefore provided with the means of life by the establishment of "Latifundia" or large scale capitalist farms using the gang labour of slaves in the territories which it over-ran of predatory necessity. Thanks to this predatory system of farming which bears so grim a resemblance to our mechanized ranch farms, large tracts of land around the Mediterranean basin which for unnumbered centuries had been among the most fertile in the world, became barren wilderness through exhaustion of the soil and consequent erosion. The Romans were great road builders because they realized that an Empire developed on this predatory principle depended upon its communications. Their great highways, striding in an arrogant straightness that holds the land in contempt, were more than the life-lines of Rome and the strategic routes of the Legions which must forever be policing its borders, they were the symbol of a fundamental change in the relationship between man and his earth.

The Roman Empire met the fate that must inevitably overtake all predatory civilizations unless they can succeed in destroying the spirit of man utterly. " Things fell apart, the centre could not hold," and after a brief and turgid history of four centuries " the glory that was Rome " collapsed with the facility of a house of cards, plunging Europe into the flux of the Dark Ages. In Europe and in England the slave farms of the Roman order were a brief interruption which vanished without trace as the threads of an older life were painfully gathered together. This was the organic life of self-sufficiency for which in man's necessity there was no alternative. Great Empires might fall and cities be put to fire and sword but still the oxen must be set to plough and the sower go forth to sow.

In this age of violence and disruption the Feudal system was evolved by which men bound themselves to a powerful leader who would protect them from aggression. Under this system the common man became a serf or bondman who undertook boon-work on his lord's land in return for the lord's protection. For this work the serf received no wages in money, nor was he free to leave the service of his lord. For this reason modernists have taught us to believe that the serf was an abject slave. They fail to point out that his boon-work entitled him to his share of real wealth in the form of his portion of the land which was his " by right of custom ", and that he was therefore unlikely to desire to leave his lord's service. It is difficult, in fact, to postulate any better system for such troublous times, and while only an idealist would suggest that it was not abused, yet arrogance on the part of the lord was held in check by the same " Custom of the Manor " which protected the rights of the common man, and which he could not infringe with impunity.

Assisted by the ideals and the unifying force of Christianity blended with the ancient wisdom of an older Paganism, order was slowly restored in Europe and for a time the unification of a continent appeared a possibility. The earlier chapters of this book have endeavoured to show how and why this attempt failed, and how, instead, man's vision of reality again became clouded by the will-to-power so that he embarked once more upon the disastrous pursuit of token wealth. We have seen also how on this occasion, thanks to the development of scientific knowledge, he has been abundantly successful in solving those physical problems which had hitherto prevented the large scale exploitation of real wealth for the sake of the power represented by money.

To-day real wealth can be defined in precisely the same terms as on the day Adam left the Garden; it is still the sum of natural resources plus human ability. Consequently its fruits can only be enjoyed when the fullest use is made of both, or in other words, by self-sufficiency. On the other hand when token wealth becomes the aim, human ability, which as we have seen, has a value which can only be assessed in terms of the real wealth which it is capable of producing, becomes, in monetary terms, a liability. For this reason the greater the extent to which human ability can be dispensed with, the greater the token profits accumulated, and as a result man concentrates always upon such natural resources as he can most readily exploit, that is to say with the least expenditure of labour. Pursued to its logical conclusion this policy eventually leads to the exhaustion of natural resources through exploitation and ignorance, to a progressive decline in the real value of the article produced, and to the complete atrophy of man's creative ability. Fortunately, however, it creates in the process a form of society so cumbersome and so unstable that its collapse before this ultimately disastrous stage is reached becomes almost inevitable.

For the effect of concentrating productive capacity and knowledge always toward that which can be produced with the greatest facility is to create, not only monopolies, but a widening gulf between producer and consumer. The wider this gap the more complex and unwieldy become the problems of distribution and of relating production with consumption, and consequently the greater the waste. Swifter and swifter methods of transport become necessary, while for every actual producer of goods an ever-growing army of men must be engaged in parasitical occupations attendant on that production, a liability which still further detracts from the real value of the article produced and from the reward of the producer. Of the problem of relating production with human need across this gulf and under these conditions of production enough has already been said. Individual human need they cannot under any circumstances satisfy. They can only meet the needs of humanity in general by imposing the acceptance of complete uniformity, and could only achieve this end by the emergence, after recurrent periods of cataclysmic warfare, of a system of world control of the means of production and distribution so rigid and so complex that the mind boggles at its conception. They produce, in fact, conditions in human society which are the direct opposite of those which result from the fullest use of natural resources plus human ability, diversity where there should be uniformity, uniformity where there should be diversity.

It might seem from this that downfall such as that which overtook the Roman experiment is inevitable. This may be so, but the object of this book is not to make dark prophecies of doom but, having greater faith in the potential greatness and eventual redemption of the human spirit, to outline a possible way of averting such a catastrophe. Manifestly there is only one way in which this can be achieved, and that is by changing the social structure from that of a society based on the acquisition of token wealth to one based on the real wealth of local resources and human ability. This means adopting the principle of maximum self-sufficiency in the individual life no less than in that of the community and the nation.

In the eyes of the "scientific progressivist" this is a defeatist notion born of rural sentimentalists which involves the abolition of all machinery and the wilful abandonment of all the accumulated knowledge of three centuries. Accordingly he regards it with undisguised contempt as unworthy of serious consideration. In fact the project does not imply the abolition of scientific knowledge or of the physical principles which it has evolved, it merely advocates their wise use as an aid to self-sufficiency by their employment as a complement to, rather than a substitute for human creative ability. This would be the aim of a generation educated after the manner suggested in the previous chapter. It is an aim that cannot be achieved easily and rapidly by arbitrary methods imposed by a minority however well intentioned. Such attempts have been made on a small scale in the past and have failed. It is no use driving a herd of unemployed urban poor into the country like so many cattle to pasture, settling them on smallholdings and leaving them to get on with it, any more than it would be efficacious, assuming it were possible, to close down all large scale mechanized industry overnight. Fish cannot be taught to walk upon dry land, and the accumulated influence of three centuries of predatory materialism cannot be undone in the space of a lifetime. Hence the emphasis upon education and practical training as the essential primary factor in any process of transition.

The influx into society of a generation whose creative ability had been fully developed and who possessed a sense of real values, might slowly but surely reverse the fatal course of all productive work towards the coalescence of specialization. In this process of "breaking down" monopoly industry they would be guided by the following principles which, as they progressed, would become increasingly self-evident:—

That mechanical methods should only be employed in work provided the qualitative result achieved is better in the eyes of both maker and user, and provided they cause no damage to, or wastage of, the natural resources upon which they operate. "Cheapness", that is to say a saving in labour by greater facility of production, should never be countenanced at the expense of this quality.

That no man should make any use of a machine or a scientific process unless he possess a comprehensive knowledge of its principle, function and purpose, and is already skilled in the use to which it is to be applied.

The first of these principles cannot be fulfilled by centralized quantity production because, although the product might possibly satisfy the producer, it cannot satisfy every individual consumer of whose particular need the producer cannot be aware. The second principle is inseparable from the idea of self-sufficiency, otherwise man is not the master of machine or process, but the slave. Both imply the disintegration of large scale industry into smaller and smaller self-contained units, a process not to be confused with the war-time "dispersal" of industry into ever more specialized groups dependent upon each other and existing only to serve a "parent" company. This diffusion of industry could not fail to be a lengthy process extending through several generations since its success depends not only upon the educational enlightenment of worker and executive alike, but upon financial and political reform. Of these the last-named will be considered in the next chapter.

So far as the financial question is concerned it should be obvious that a close association between real wealth and token wealth must be restored. Money, like the machine, should be the servant of the producer, not the master. This end can only be achieved by repealing the Tonnage Act of 1694, and so returning to the State the responsibility for the issue of currency commensurate with the nation's creation of real wealth, instead of by financial houses as a debt "created" on paper, interest on which can only be paid by exploiting both man-power and resources. By issuing money upon this credit basis, by limiting interest on capital, or, better still, by making interest free loans to local communities in order to assist them to make the best use of their local sources of real wealth, the State exercises its proper paternal function which is in no sense an arbitrary imposition. The interest on such loans consists of the value to the community of the real wealth produced.

The further the policy of diffusion in the interests of self-sufficiency is carried the more clearly it will become apparent that the industry which must form the basis of the whole social structure is agriculture. It is upon the earth and upon the labour of the husbandman that the livelihood of all men on this planet depends, and therefore the town and the industries of the town must resume their proper function as the servants of the countryman. Only a civilization based upon falsehood can imagine that the country exists to serve the town and that both the countryman and his fields can be exploited or neglected according to the whim of urban bureaucrats and economists. A tragic separatism symbolized by the can-opener and the bottle-on-the-doorstep has made the peasant a fiction of urban sentimentalists and the rural craftsman a museum piece. It has begotten the pernicious idea of the land as something to be "conquered" for token profit by the saurian machines and chemicals of the mechanized ranch-farm. The great American dust bowl affords only one instance of the incalculable loss of real wealth which the world has sustained as a result of the pursuit of this disastrous policy with its monoculture, deforestation, and consequent soil erosion through loss of fertility. A generation educated by the historical methods suggested in the previous chapter could not fail to regard the land, not as a potential source of profit, but as a way of life, and by so doing re-affirm the only basis of a stable social order.

In the past, owing to the lack of communications, man was compelled to live in accordance with the principle of self-sufficiency, and consequently the fact that his well-being was wholly dependent on the earth's fruitfulness aided by his own labour was self-evident. Yet, by reason of the same physical limitation he was unable to overcome inequalities caused by climatic and geological variations over the earth's surface. In consequence, as we have seen, this inequality gave rise to recurrent conflict for possession of the most fertile regions. Even upon a small scale within the confines of a small area one community might enjoy plenty while another but a few miles distant was reduced to the point of starvation by the failure of a crop. There was thus much hardship, suffering and inevitable discord. Were we voluntarily to re-adopt the principle of self-sufficiency to-day, however, we could overcome this natural inequality; this is the proper function of that modern knowledge which we now so grossly misapply. While individuals and communities in less favoured territories could, in the light of our knowledge, achieve a real prosperity and a degree of self-sufficiency without hardship undreamt of in the past, it should be the purpose of modern transport to smooth out such inequalities as still existed by ensuring the rapid interchange of surplus goods on a non-profitmaking basis. In this way the surplus fruits of an abundant harvest in one country, for example, would be directed to meet the needs of another less fortunate, and the economic fallacy of "over-production" exploded.

Over-production has no reality until all human need for the article produced is satisfied, and the further the aim of self-sufficiency is pursued the easier becomes the problem of equating production with need on a world-wide scale, and the less the burden on transport. This diminution of the need for transport would of itself be beneficial since it constitutes unproductive labour the cost of which must be borne by the consumer who thus pays a price in excess of real value for the goods transported. By thus aiming to reduce the parasitical intermediaries between producer and consumer, and, by regulating the issue of money in proportion to real wealth produced, the other economic "problem" of purchasing power would also be solved. That this problem of wages and prices can never be solved so long as paper money is issued as an unredeemable interest bearing loan is certain. For who can be so blinded by the conceit of modernism as to defend a system which demands the gross exploitation of labour and resources in order to pay the interest on its fictitious loans, and, having done so, orders the destruction of the article produced because it would "spoil the market"? How justify a civilization which "profits" by scarcity rather than by abundance, which compels the fisherman to return his catch to the sea and the farmer to burn his crops when the bellies of the poor are empty?

One question the modernist will inevitably pose will be the practical one of how our scientific knowledge can be reconciled with and applied to the principles which have been outlined. It will be argued that many of the machines and processes which we have evolved and which we have come to regard as essential amenities of civilized life, necessarily involve a system of large scale mechanized production. Once we can rid our minds of the fallacy of cheapness measured in terms of the false values of token wealth, however, and grasp the qualitative conception of real wealth, we shall find that there are no modern machines or processes of real or potential value to the human race which do not lend themselves, or which cannot be adapted, to small scale qualitative production for local need. At the same time many of the more complex machines and processes would be automatically ruled out because they are not adaptable to individual needs, but can only produce a stereotyped and inferior product, or are wasteful of natural resources. Again, the demand for others would be reduced to an extent which would make small scale regional and qualitative production practicable. In this latter instance the machinery of transport, the aeroplane, the motor car, the steamship and the locomotive may be cited as examples. These machines would obviously be used to a reduced extent as self-sufficiency increased, yet no one can deny their potential value as a legitimate means of overcoming natural inequality, and as an aid to international understanding and goodwill.

To attempt to outline the method of adaptation of every branch of human activity and scientific knowledge to the principle of self-sufficiency would manifestly be beyond the scope of this or any other single volume. It is also unnecessary, because once the ends had been determined, the physical means would shape themselves. It is only possible to state general principles and use examples as illustration. Electricity, for instance, which upon first consideration might appear to be incompatible, is capable if properly applied of performing work of incalculable service to man. It can render the factory obsolete. It represents the most effective means yet discovered whereby local sources of power could be utilized and distributed to the benefit of the local community. In some districts water power, which our forefathers harnessed to such good purpose, in others coal, and in others charcoal or waste products (used in conjunction with the gas producer and gas engine), all these could be utilized as prime generators. Even such ancient principles as the tide-mill, or such modern ones as the possibility of using sewer gas, should not be neglected. To develop all these regional resources to fulfil regional needs is to practise true economy and to contribute to the real wealth of the world. It is futile to argue that it is more economical to neglect them as we have done, concentrating instead on huge and costly power stations relying either upon fuel transported long distances to the site, or upon water power secured by damming and flooding fertile valleys. This is exploitation, it is not economy. Further, such centralized power schemes necessitate hundreds of miles of costly power-lines with their attendant booster or transformer stations which occupy more valuable land. From every point of view, including the strictly technical one, such long distance distribution is uneconomic, and by the time the current is brought to the consumer it can never be "cheap"; if it appear to be so then it is dearly bought by exploitation at the source.[1] Yet still we perpetuate our folly, the Highland Electrification Scheme being the latest iniquitous example of a plan framed to exploit the Highlands of Scotland for the benefit of monopoly industry without the slightest consideration for local interests and needs.

To a self-sufficient community the availability of really cheap current from a local source could be of inestimable value, the small electric motor, reliable, odourless, noiseless and adaptable being potentially an admirable servant of man; a hewer of wood and drawer of water, an aid to husbandman and craftsman. This is the true function of science and the machine, and what is true of the example of electricity chosen here is true of other applications of modern knowledge which we now misuse. We can never know contentment or enjoy real wealth unless we grant man the freedom of his ability, and we cannot do this until we learn to employ science in this way as a complement to, not a substitute for, the skill of hand and eye, and as a means of developing all our sources of real wealth, instead of as a method of wholesale exploitation for token profit. Only in this way can we re-establish those vital associations which enable the individual to become aware of his place in time and space and so make possible the good life.

[1] Even when no account is taken of distribution losses, the average power station using coal is only 20 per cent. efficient. 55 per cent. of the coal energy is dissipated in the cooling water.

Chapter XIII

SOCIAL AND POLITICAL ORGANIZATION

So far we have considered the spiritual desires of man, how they may develop for good or for evil, and how, by a system of education we may encourage the good and check the perversion of evil. We have argued that the principles of that education can only be applied in the life and work of the individual in a society whose goal is self-sufficiency. Having progressed so far we now have to consider the possible social and political organization of such a society.

In our modern urbanized world, industry, or in other words, the factory, has become the basic social unit. The cities in which the vast majority of our population dwell are not the product of any communal unity of purpose, possess no actual corporate identity, and therefore cannot be considered as integral social units either in whole or in part. They simply represent huge agglomerations of population grouped about the nuclei of the factories. The inevitable coalescence of the machinery of profit-production into fewer and larger units is thus automatically accompanied by an ever increasing concentration of population. The greater this concentration the more unbalanced the distribution of population over the earth's surface, and the more unstable the society. Unstable because the more monopolistic and specialized industry becomes, the more the inhabitants of the surrounding "wen" become dependent for their livelihood on forces beyond their individual or communal power to control. Thus specialization promotes social conditions the direct opposite of those created by self-sufficiency which necessarily implies industrial diffusion. This exposes the fallacy inherent in one of the most popular modern political theories which argues that the modern factory, like the village of the past which it has replaced as a social unit, should become a unit of regional self-government. The village, however, possessed an independent and organic life of its own in which every aspect of the life and work of the individual and the family was intimately related to that of the community.

Responsible self-government was thus an integral and inseparable part of the village structure which eventually collapsed, not through inherent fault, but by reason of the pressure of hostile external forces which destroyed the whole edifice. Self-government, if the term is to possess any real meaning, necessarily implies individual responsibility, and this in turn is synonymous with self-sufficiency. It is therefore clearly absurd to imagine that such a principle is applicable to an organization based on the dissolution of responsibility and the specialization of function. Only the technocratic "queen-cell" is capable of governing the factory and then only in a limited sphere, because a monopoly organization is necessarily subject to centralized control bearing no relation to local needs.

The previous chapter made clear that because the land and the fruits of the land are the prime source of real wealth, and therefore the factor upon which the life of mankind is dependent, it follows that a stable and self-sufficient society can only be built upon an agricultural basis. The life of every soul upon this planet depends upon the earth's fruitfulness, and therefore the foundation stone of that society must be the husbandman who labours to augment that fruitfulness. This is the primary and inescapable craft of man about which all his other activities should revolve as complementary satellites about their "bright particular star." This fact should be a platitude, but in a world devoted to the pursuit of token "profits" those few who make so bold as to declare it are but voices crying in the wilderness. So complete is the inversion of true values in the modern world that of all the occupations of man that of husbandry is considered the humblest and most menial, and is consequently the most exploited and the least rewarded.

Profit margins, the interest on paper loans and the chain of *entrepreneurs* between producer and consumer all demand their pound of flesh, and in consequence, as we have seen, in a civilization pursuing these profit motives human labour becomes a liability the cost of which must be minimized in the attempt to reconcile prices with purchasing power. Manifestly the only way in which this reconciliation can be truly secured is by encouraging individual production of real wealth, by increasing the proportion of the population engaged in that production, and by ensuring them a just reward for the real wealth they produce. Where the actual producer must bear the incubus of the profit margin, financial interest, transport and other unproductive costs, he can never receive such reward, and therefore the gap between prices and purchasing power cannot be closed. A vicious circle is thus created which cannot be broken by any policy, however revolutionary it may appear which pre-supposes the continuance of the present economic system. If wages and returns to the producer are reduced in order to lower prices, then the purchasing power of wage earner or producer is proportionately lowered, and therefore no improvement is secured. Conversely, if wages and returns are increased, prices increase, so that the value of currency falls and higher wages represent no increase of purchasing power. The only solution of this problem which the modernist has been able to evolve is to eliminate the liability of human labour as far as possible. If machines can be introduced which will enable the work of ten men to be carried out by one, then the cost of production will be maintained at a low level, while the wages of the remaining worker, because of his increased productive capacity, can be increased, thus affording him adequate purchasing power. It has been the purpose of this book to expose the utter falsity and folly of this policy, and to show that it succeeds in bridging the wage-price gulf only at the cost of a progressive decline in the real value of the article produced owing to the increase of unproductive costs, and the total loss of any relationship between production and individual need. Furthermore, similar employment must be found for the nine men replaced if they are to receive adequate purchasing power, and this entails further industrial expansion whether or not there already exists any need for such expansion. Thus this progressivist policy can never achieve a stable phase, but is dependent for its success upon continual unregulated and "immutable" expansion which accelerates the more vigorously it is pursued. We have already tasted the bitter fruits of this expansion in two world wars, so that it should be unnecessary here to follow it further on its course toward the ultimate nightmare destination of world chaos to which it inevitably leads. Any check to this process of expansion at once produces the symptoms of economic disease with which we became so familiar in the period between the two world wars; unemployment, "over-production", inflation, depressed currency, and alternating booms and slumps.

What then is the position of the peasant, the primary producer who should be the foundation stone of our society? In this country, because we pioneered the industrial system, he has been eliminated by a process already outlined. In the more " backward " countries of Europe and Asia, however, he still survives. Survival is the operative word since he is engaged in a struggle for bare existence which becomes increasingly bitter as the world of " development " in which he finds himself becomes ever more hostile to the way of life of which he is the exemplar. The intensity of that struggle naturally varies from country to country according to fertility and climatic conditions, but everywhere he shoulders the same adverse economic incubus. In regions more favoured by nature in these respects he may, even without the facilities which this incubus denies him, achieve a degree of self-sufficiency which renders him to a limited extent independent of economic vagaries. He may not prosper in this way, but at least he is not haunted by the fear of starvation. Other countries, in an endeavour to assist him, have imposed artificial restrictions on the importation of " cheap " food. In this way they have ensured him a fairer price for his produce, but at the expense of the purchasing power of the landless worker in the area over which the tariff operates. Industrial expansion must therefore be encouraged in the attempt to overcome this deficiency. In many regions of central and eastern Europe, however, the lot of the peasant, never better than one of bare subsistence, has been reduced to starvation point. It may be noted also that wherever in these regions conditions of large scale farming by landless labour obtain, manual labour is used *in preference to machines* provided wages can be maintained just above starvation level. So soon as wages rise to a more humane level, machinery is introduced and the " surplus " labour driven off the land.

Obviously there is only one true and permanent solution of the peasant problem if it is viewed in the light of the reality of absolute values. The food producer must receive the just price of his labour and be encouraged, by interest free loans to develop the fruitfulness of his soil and so to contribute to the real wealth of all mankind. The greater that fruitfulness the greater the population that can be occupied in and supported by agriculture, and the greater the degree of self-sufficiency. Such development involves the free application of our modern knowledge, not for token profit, but to secure this real wealth. It involves schemes of irrigation and reclamation, and the selective breeding of plants and livestock more adaptable to variations of soil and climate. By such means the principle of self-sufficiency can spread over the earth's surface, gradually overcoming natural inequality and making the "desert to blossom as the rose". Furthermore, it implies the adaptation of manufacturing industry to satisfy the needs of the countryman rather than the fatal inversion which we perceive to-day. Instead of an industry becoming increasingly centralized and specialized, this pre-supposes an industry splitting into smaller and smaller units and becoming ever more diverse in its local activities as the scope and range of local agriculture increases. In this way we can approach the ideal end, which is by no means Utopian, where the interaction between industry and agriculture becomes so close that they become part of a local self-sufficient whole imparting to life a pattern of common purpose. The industrial worker would then no longer be divorced from the realities of life, nor would he be denied the freedom of his ability. In all probability he would himself be a part-time agriculturist, either possessing a holding of his own, or devoting himself to his craft at certain periods of the year, and to agriculture during its periods of greatest seasonal activity.

Is this the solution of the urban agronome to the European peasant problem? Obviously it cannot be because he pre-supposes that present economic laws are immutable. One of the latest contributions to this subject[1] advocates the usual progressivist nostrum. It concludes that the present peasant population is excessive, and that the surplus should be combed out and absorbed by inaugurating new industrial development. Machinery and monoculture should then be introduced in order to raise the "output-per-man" of the remaining peasantry. In this way, the argument runs, a stable class of petty-capitalist peasants would emerge. In fact no such stability could be achieved because, as we have seen, once this disastrous policy is launched it is automatically expansive, calling for increasing mechanization, proportionate de-population, and consequently the eventual conversion of Europe into a wilderness of Steppes, and mechanized ranches.

Local and regional self-government only becomes an attainable and practicable ideal in a self-sufficient community whose every activity is closely inter-related and which has as its basic unit the family and the agricultural settlement. The greater the degree of self-sufficiency attained within the region, the greater the responsi-

[1] *Food and Farming in Post-War Europe*—Yates & Warriner.

bility which may devolve upon the local government. Manifestly, however, there must always exist a wider system of social and political organization not only upon a national but on an international scale. The higher the standard of self-sufficiency and therefore the greater the local responsibility, the simpler the machinery of this wider organization.

In this country we already possess machinery of local government; we have parish, rural, urban and county councils, and at the head a national government consisting of the elected representatives of the people. If, however, we compare this political structure with a tree, we perceive that its true function is inverted. Instead of roots nourishing and supporting the trunk, local government represents branches of that central trunk. So far from conserving and renewing the sap of political vitality they are thus the first to wither and fall in the winter of adversity. Therefore, instead of representing the corporate voice of the region, our local government bodies have become to an increasing extent, the instruments of centralized control. Economic dependence is necessarily accompanied by a similar social and political trend. Here it is possible that the modernist will argue that, local government apart, our system of parliamentary election guarantees the health of the body politic by fulfilling the democratic ideal of government of the people, by the people, and for the people. Certainly no one can cavil at this ideal, but the question is whether our election system truly fulfils it. The actual choice of candidature for election rests with the respective party offices. Of the real motives actuating these parties the common man knows nothing; he is as ignorant of the character and purpose of the candidate proposed and of the reason for his proposal as the candidate is ignorant of the electors' particular needs. The electors' choice is thus arbitrary and is largely conditioned by intensive party propaganda. Consequently government is entrusted (to use Cobbett's paraphrase of Voltaire) to "Representatives of the people of whom the people know nothing". Such a system is wax in the hands of those pursuing economic power, and local government survives merely as machinery to impose their policy, the hollow simulacrum of regional self-government.

Even if such a centralized government does not consciously abuse its power it cannot ensure individual justice and welfare because it is ignorant of local conditions and needs, and must therefore impose an arbitrary conformity, based on the lowest common denominator. This is certainly not in accordance with the professed ideal of democratic government, in fact it is directly opposed to it. Government by the people can only become a reality instead of an abstract fiction if the powers of local government are progressively extended instead of contracted. This end can only be accomplished by building a reformed political structure upon the basis of the smallest unit. This basic unit might consist of a reconstituted parish council, rural or urban, whose members would be elected by vote of the adult members of every family within the parish. From these members so elected it would then be their duty to elect one or more as their representatives in a rural or urban district group. Again it would be the responsibility of every individual member of this district unit to elect one of the members of their group to a regional council. Thus, through a progressively widening electorate, national and eventually continental and world groups could be elected and formed. In this way affairs of national and international moment would be entrusted to a comparatively small council of men elected by plebiscite whose ability and integrity had already been tried and proved in the lower units of government. Any group would be empowered to act independently of the groups *above* within the scope of the powers devolved upon it, but any measures it might so introduce would only become law when they had been ratified after debate by a majority vote of all the basic parish groups in the area affected.

It is manifestly impossible to compress the design of a new political constitution into the compass of a book which is concerned only with general principles. The foregoing is merely intended as a tentative and provocative sketch of a possible system of government by inter-related groups which would have been unworkable a century ago, but which modern methods of communication have made feasible. Let us suppose that such a form of group government has in fact been constituted so that we may imagine its operation and effect.

Introduced in our world as we know it to-day the great burden of responsibility would naturally fall upon the national and international groups. Theirs would be the task of stabilizing and controlling the issue of currency commensurate with the output of real wealth and in the form of interest free loans. They would also incorporate within their framework an international and national marketing organization operating on a non-profit making (lease-lend) basis, and to oil the wheels of this machine it would be the particular task of the international groups to determine world standards of currency, weight and measurement (preferably on the metric system) and a simple basic language of commerce. This marketing organization covering the whole of the group system would constitute the channel through which production could be reconciled with need and national and regional inequalities overcome. In this way also the swollen bulk of mechanized profit-production industry would be used, so far as it is capable of such adaptation, to provide the initial facilities which would enable the foundations of local self-sufficiency to be laid and local resources to be developed. In the course of such a process centralized industry would inevitably seed itself like an overblown poppy-head, while growing local resources would be in a position to absorb and rehabilitate the population released from the industrial cities. This in turn would mean that the task of national and international organization of distribution would become progressively simpler so that the "seeding" of industry would be accompanied by a similar "seeding" of the responsibility of government into the regional and district groups. Obviously the higher government groups could never be dispensed with, but, as local self-sufficiency grew, they would become to an increasing extent servants, or at the most wise parents, rather than masters. Only when that stage is reached shall we be able to say with truth that we have realized the democratic ideal of government for and by the people.

Consider the implications of this. The State would be the only landlord and landowner, and the peasants or small proprietors and their heirs or nominated successors, would hold their property from the State "by right of custom" and upon trust of their good husbandry and good workmanship. If the State is represented by a centralized body, then this ownership and the power of eviction which it carries would be an imposition of intolerable bureaucratic tyranny and State absolutism. But under group government the State, for the individual, would be represented by a local group well known to him, conversant with his needs, and by his own will elected. In exercising such powers this group, like the mediæval village court, would thus impose no arbitrary control from without. Similarly, the State should be responsible for the sick, the aged and the needy, but if such a charge is administered by a central authority with no matter what conscientious attempts at justice, it can give no more than an impersonal and dispassionate dole, the receipt of which brings fear, shame or demoralization to the recipient. When, on the other hand, the "State" is a local body its particular care of the needy does not destroy the personal element, and has the compassion of true charity. Thus it will be perceived that self-sufficiency in government is no less important than, and in fact is inseparable from, the principle of self-sufficiency in work.

The second part of this book was concerned to describe and emphasize the "separateness" of modern urban life and the insuperable problem of reconciling within its framework the needs of the individual with those of the community. It was this very quality of "separateness" which made possible distinct analyses of modern life at work, at home, and at leisure. By contrast it is hoped that the last three chapters will have shown that once life becomes informed by a common principle such arbitrary distinctions can no longer be drawn since all these aspects of life tend to converge toward a whole indivisible and organic. In conclusion we must consider the potential fruits of that whole and its deeper spiritual implications.

Chapter XIV

ART AND RELIGION

"For us the windes do blow,
The earth doth rest, heav'n move, and fountains flow;
Nothing we see but means our good,
As our delight or as our treasure;
The whole is either our cupboard of food
Or cabinet of pleasure.

The starres have us to bed,
Night draws the curtain, which the sunne withdraws;
Musick and light attend our head,
All things unto our flesh are kinde
In their descent and being; to our minde
In their ascent and cause.

Each thing is full of duty:
Waters united are our navigation;
Distinguished, our habitation;
Below, our drink; above, our meat;
Both are our cleanlinesse. Hath one such beautie?
Then how are all things neat!

More servants wait on Man
Than he'll take notice of: in ev'ry path
He treads down that which doth befriend him
When sicknesse makes him pale and wan.
Oh mightie love! Man is one world, and hath
Another to attend him."

George Herbert (1593-1632.)

Before the Modernist arrogantly dismisses this book, let him ask himself soberly whether it is in fact as Utopian as the dream of the leisure state which he cherishes. This dream is one of a World State in which machines will have emancipated man from the slavery of labour and will thus enable him to lead a life of ease and "culture" hitherto undreamed of. The most ardent progressivist can envisage no higher end that this; indeed, it is the only possible one which can be reconciled with his belief in automatic progress. But such an end, even if it were achieved, would be barren, because the only positive freedom of man is the freedom of ability which enables him to use his creative instinct, and this freedom the machine, when used as a substitute for human labour, denies. In these circumstances man is not emancipated by the machine but enslaved, while freedom can possess only negative values which bring to leisure only boredom and decadence. "Culture" it cannot bring since all the arts of man are the product of his ability, and therefore if he is denied the freedom of that ability there can be no art. If the modernist replies to this by arguing that in the Leisure State man would have ample time to follow creative pursuits he at once implies a limitation of the use of the machine, and by so doing refutes his own theory of automatic progress. For, under the present dispensation, the machine, as we have seen, cannot be arbitrarily confined to the four walls of the workshop, but invades every walk of life, so that no such distinction can be drawn between work and leisure.

If then we accept the theory of automatic progress this Leisure State becomes the only conceivable end. Similarly, if we adopt the freedom of human ability as our aim, self-sufficiency inevitably becomes the end envisaged. This is our choice of purpose. Self-sufficiency seeks only to eliminate the unproductive and uncreative tasks which are the proper function of the machine, while positively its aim is to restore to human labour that freedom and dignity which are essential to the production of real wealth, and without which the arts of man perish. It postulates no rosy, fabulous existence of unlimited leisure from which all toil has been banished, but a world in which men will still know sweat, hardship and sorrow, yet will not be deprived of their just reward of rest and contentment. For life with all its values is dependent upon the eternal association of opposites, and therefore leisure without toil can have no value and is emptied of content. The laden table and the fire's warmth can have no meaning for those who have not hungered or felt the bite of frost, the evening's cool no benison for him who has not laboured in the drought of noon, joy no sweetness for those who have not sorrowed. This recognition of life as a pattern of light and shadow implies no fatalistic acceptance of evil, pain, and suffering as inevitable. Man must labour always to alleviate these things, and the greater his labour the more triumphant his transcendence, because the shadow of death, the tragedy of immanence, rests eternally upon all things and is thus the final arbiter which gives life savour and purpose.

So much for ends. But when, as in this book, we attempt to gather together in the present the threads of past and future, and by so doing to consider the destiny of man, the practical value of our efforts is judged, not so much by the quality of the end proposed, as by the feasibility of the means whereby that end can be achieved. If there already existed any common basis of human understanding and purpose compatible with the end in view, then means would shape themselves, but, since this is not so, a consideration of means is vital before the end can become acceptable. In the previous chapters we have considered the means by which we might approach the principle of self-sufficiency, and shown that the way is long and strewn with difficulties. But is that way any more difficult than that which leads us to the goal of the leisure state, and may it not prove to be less bloody? For the leisure state pre-supposes no less than an all-powerful world control of the means of production, or in other words of resources and man-power. If we accept for a moment the modernist's theory of automatic progress on present lines we see that it must be the function of this control to allocate to different nations world monopolies for those commodities which they can mass-produce most "cheaply". On the wastefulness of such a system, the rigidity of the control necessary and the soulless uniformity it must perforce impose there is surely no further need to dwell. But who is to enforce this control and how is the allocation of the respective world monopolies to individually predatory states to be determined except by the bloody arbitrament of further and still more catastrophic warfare? Such a form of world control represents the only alternative to chaos which automatic progress can postulate. Does this end justify the means? Let the modernist ask himself these questions before dismissing as idealistic the possibility of an alternative.

In the mechanistic world of materialism neither art nor religion can live except as embalmed survivals. For art is the product of human ability operating in qualitative freedom upon its chosen medium with no ends in view but perfection and fitness for purpose. No matter how far the result may fall short of the end desired, it yet achieves beauty because beauty is the inevitable flower of the truth of this organic association between mind and matter whereby man fulfils his natural function. All that is not the product of such an association is unnatural and false and cannot therefore achieve beauty. In a world which prohibits the freedom of ability and squanders the material upon which it feeds, it is therefore useless to prate of "culture" and the possibility of artistic renaissance. A ready-made æsthetic whether modern or traditional cannot be arbitrarily imposed upon a society in which the principles of art do not exist. There can be nothing more fallacious than the modernist's assumption that "culture" can be preserved as a vital influence so long as a few urban cliques of artists and "high-brows" are permitted to survive like so many carefully tended hot-house plants, or that the values of art are recognised and furthered if an industrialist has his wares designed in Chelsea, Bloomsbury or Broadway. This, like so many other aspects of our modern thought, represents a fatal inversion of real values. The fine arts are a flower whose roots are what we call the applied arts, which in the widest sense mean the everyday activities of man in the course of his labour. If these roots flourish the plant will flourish and blossom in profusion, if they perish, the flower, no matter how carefully it be tended, will wither and fall, the flower being dependent on the root, not the root upon the flower. A building designed by an architect of vision may be better than one thrown up by a jerry-builder to the drawings of a hack draughtsman, but unless the architect shares with the workman the same qualitative conception, knowledge of materials, purpose and vision, the building will be false and cannot therefore be beautiful. The beauty that results from the truth of perfect fitness for purpose can only be achieved if that purpose be to satisfy local human need to the best of human ability, and if fitness implies the fullest use of local resources. Thus the dicta of the "functionalist" school, "Form follows Function" and "Beauty is Fitness expressed", are dangerous half truths lacking definition, while Le Corbusier's oft quoted statement *"la maison est une machine a habiter"* sums up the modernist's soulless and desolating ignorance of man's spiritual needs. Almost alone among modern architects, Frank Lloyd Wright has recognized that man needs something more than steel and concrete, and speaks truth when he writes: "The first condition of homeliness, as it seems to me, is that any building which is built should love the ground on which it stands."[1]

What is true of architecture applies with equal force to all other aspects of art. The relegation of the artist to the position of a cultural specialist and a member of an exclusive group is as damaging to himself as it is to his art. While the exponent of the applied arts becomes obsessed with the impossible task of evolving some formula which can be reconciled with modernism, the fine artist seeks shelter in some introspective clique, evincing that arid intellectualism without tradition and without purpose which becomes ever more eclectic and emasculated as it adds abstraction to abstraction and obscurity to obscurity. The unpalatable truth is that modernism is fundamentally hostile to art, and that its acceptance puts the artist in a false and untenable position where art can be no match for the philistine.

Only by pursuing the aim of self-sufficiency can we move toward that freedom of ability by which alone the humblest work of the common man can be infused with that truth and fitness which is beauty. Only out of the richness of such an organic life of infinite variety allied to common purpose can we look for a true renaissance of the arts of man. For no matter how deep the artist's perception of absolute truth may be, he cannot express his vision except in terms of the relative values of contemporary life, and if these values have themselves no basis in that truth he can only portray their discord, while he can never capture the mind of the people. But in the fertile soil of organic life the roots of the arts, or in other words craftsmanship, flourish, and from them blossom, in inevitable consequence of regional and seasonal festival, in custom and ceremony which represent the grace and crown of labour, poetry, music, painting, drama and dance. This was true of the past and will be true in the future so soon as we choose to prepare the soil.

Lest there be any who doubt, despite all the evidence to the contrary, that the principles of art ever did, in fact, inform the life of the common man, but imagine that this is a romantic fiction, here is an illustration which survived in country memory until twenty years ago. It is a song once sung by the sowers of Wiltshire and Gloucestershire as they set out upon their immemorial task:[2]

[1] *An Organic Architecture: The Architecture of Democracy.*
[2] From Lechlade. *Folk Songs of the Upper Thames.* Collected by A. Williams (1923).

"Now hands to seed-sheet boys,
We step and we cast. Old Time's on wing;
And, would you partake of harvest's joys,
The corn must be sown in spring.

Fall gently and still, good corn,
Lie warm in your earthy bed;
And stand so yellow some morn,
For beast and man must be fed.

Old earth is a pleasure to see
In sunshiny cloak of red and green:
The furrow lies fresh, and this year will be
As years that are past have been.

Old mother, receive this corn,
The son of six thousand golden sires;
All these on thy kindly breast were born,
One more thy poor child requires.

Now, steady and sure again,
And measure of stroke and step we keep:
Thus up and down we cast our grain,
Sow well, and you shall gladly reap."

This is an example of past riches that, like a rose blooming out of season, has survived into the bleak November of our day. Thanks to the labours of the collector it has been preserved on the printed page where it remains to remind us of lost grace. It can do no more for it is as dead as a stuffed bird in the glass case of a museum, because the way of life of which it is the logical expression has perished. It did not perish without struggle, having endured through countless vicissitudes since "time out of mind", but we in our time have seen the eyes close perhaps forever. The last generation in whom the memory still lived are gone, and their children have only the songs of the cinema on their lips.

Consider this song. Essentially realistic since it has to do with the everyday life and work of the fields, simple—to us almost naïve—in its language, it yet conveys an inner depth of meaning, as profound as Traherne's "Orient and Immortal Wheat," which is the very stuff of poetry. It expresses a tacit recognition of the dependency of man upon the earth's fruitfulness and upon his own labour to increase that fruitfulness. It also reveals the sense of continuity, the association of immanence with transcendence, which informs the sower in his immemorial task; he knows that his feet follow in the wake of those who, with the same wide gesture, cast the precious seed six thousand years ago, yet this awareness breeds, not that arrogance which seeks to "conquer" nature, but a humility which is religious in the deepest and truest meaning of the word. Thus in this simple song of men whom the modernist would call "primitive" and "illiterate" we come to the fountainhead of art where it is at one with labour and with religion.

From the day when man first discovered his unique power to create, and realized that by employing this gift in harmony with natural forces and processes he could increase the earth's fruitfulness, he began to wonder at their ordering, and to ponder upon the reason for that order. He did not only ask himself how, but also why. Why did the small dry seed he learned to plant sprout and multiply? Why did the earth nourish it, the moisture of dew and rain and the sun's warmth encourage it? As his knowledge and ability increased, his experience of cause and effect enabled him to determine *how* these things happened with increasing certainty, but the order and subtlety which was thus revealed to him led him to ponder yet more deeply the problem of first causes which has exercised the mind of man through all the ages. His labour as a husbandman kept him constantly in intimate touch with the recurrent but eternal mystery of decay and renewal, a pattern so ordered that he could not fail to believe it to be purposive, and therefore the work of some sentient creative power possessing a timeless omniscience beyond the range of his understanding. Thus it was that he came to regard his environment and his work, not merely as the result of a dual partnership between man and nature, but as the product of a trinity whose first member was this mysterious creative power which was at once the author of all. Such a power and such an authorship being completely beyond the limited range of human understanding, man found it necessary to attribute it to some representational symbol before he could express his belief in coherent form. At first he invested with godhead one or more of those natural forms which appeared to him to be the most significant, life-giving, remote or mysterious. The phallus he worshipped as the emblem of fertility and of eternal renewal, or the sun as the giver of all life. But it was natural that as man became aware of his own unique creative powers he should attribute godhead to a creator of superhuman form. Sometimes he invested one omnipotent being with the authorship of the universe, at others he evolved complex heirarchies of gods some benevolent, other malevolent. The natural world was often believed to be a battleground of warring gods no less than the world of men, and these deities, easily displeased, had frequently to be placated by human sacrifice, while the priesthood who constituted their temporal ambassadors wielded in consequence an all-powerful weapon of superstitious terror which they often employed to their own advantage. This conception of the jealous tribal god has given rise to the modernist's theory of the origin of religious belief popularized by H. G. Wells. This traces the belief in godhead to the predatory "Old Man", the symbol and figurehead of a warlike tribe. This supposition rests, however, upon the theory that man has always been by nature warlike and predatory, whereas recent research has given us very good grounds for the belief that there was a period in human history when warfare was unknown. At one stroke this not only transfers the legends of a golden age from the abstract future to the historical past, but destroys the modernist's argument.

It is safe to say that so long as man leads an organic life in harmonious association with the natural world he perceives himself to be a part of an all embracing harmony the order, beauty and subtlety of which he logically attributes to a supreme intelligence. This perception induces those feelings of wonder and humility which form the essential basis of religion. Any religion not founded upon, or deprived of, this fundamental attitude of mind is false. But because of man's unique faculty of free-will, his organic association with his world could only be maintained by deliberate choice on his part, for if it were instinctive and automatic he would be deprived of his great potentiality. This potentiality, this creative gift, made him not merely a passive participant in a universal pattern, but an active contributor, or, if he chose to misuse it, a disruptor and destroyer. It was precisely through the misuse of this gift in the pursuit of power that man became a warlike and predatory animal. By so doing he became arrogant and proud, this pride increasing in proportion as his knowledge increased, with the result that he lost his sense of wonder and humility. With this loss went his perception of a universal harmony, and consequently there was a fundamental change in the relationship between man and his world. Having broken the organic links between them, and so created discord, he could perceive only discord. His warlike tribal gods peopling a satanic, hostile, or predatory universe were thus the symbol of this tragic distortion of vision. In fact, this savage world of warring gods does not differ fundamentally from the puritan's view of nature and a jealous god, or from the mechanistic world of modern scientific materialism. We have simply replaced the ancient myths and deities by scientific abstractions to which we subscribe an equally implicit belief.

The Christian philosophy was simply an eloquent affirmation of the organic relationship between God, man and nature, and the story of the decay of western Christendom is the story of the severance of that relationship. Consequently the simple, yet profound principles of Christianity lost their significance for men to whom knowledge spelt power. Instead of a part of life, the principles of Christianity were divorced from the affairs of men, and religion became a code of personal morality, the church offering, as it were, an insurance policy for immortality in return for the premium of ritualistic observance. To-day the church is losing the power even of this questionable influence, for the representative of religion, no matter what his denomination or creed, who tacitly accepts the material world of the modernists, is placed, like the artist, and for the same reason, in a position which is false and untenable. For if he were to apply sincerely the principles he professed to uphold he could not do other than denounce our civilization as roundly as any anarchist, but with more authority. So long as such a declaration of principles is withheld, so long as the church is content to preach dogma instead of fearlessly proclaiming a realistic doctrine, so long will the majority remain irreligious, and the ceremony and ritual of religion be empty of significance. For these latter, like art, constitute man's attempt to give expression to his deepest perception, and if that vision and belief fail him they become meaningless. It may be that not only art but religion, in the ritualistic sense, must die before it can be born again; that only after the re-adoption of an organic life out of dire necessity shall we recover that sense of wonder and humility without which neither can flourish.

The religious view of life rests upon the belief that the universe is purposive, and that its ordering is not therefore the result of the mindless interaction of inanimate forces, but a work of conscious deliberation. While our knowledge does not enable us either to prove or to disprove this assumption, its rejection has a more slender basis in reason than acceptance. In fact the more deeply we probe into the problem of first causes, into the intricacies of natural processes and the origin of the human mind with its power of free will, the less feasible does the theory of the mechanistic universe become and the more logical the religious view. Pursued to its ultimate conclusion the materialistic theory ends in nihilism, in a world blind and purposeless, signifying nothing, leading mankind, a chance combination of inanimate forces, out of nowhere into nowhere. Obviously in such a world there can be no standards or principles except of man's own devising, and, since he is a predatory animal caught up in an eternal conflict of opposing forces, such principles must necessarily be governed by expediency and therefore be subject to change. If the natural world is thus considered as an unending storm which man must ride on his passage through life, then his ability to conquer and master it becomes the measure of his success, and consequently, no matter how many abstract humanistic principles he may try to evolve, might remains ultimately the only certain right. Yet just as nature abhors a vacuum so the human mind cannot accept so empty and bleak a view. It seeks refuge in innumerable abstractions, scientific, intellectual, or esoteric, whose only common denominator is the Utopian belief in man's ultimate perfectibility. Such a belief has no basis in reason whatever. It cannot even be clearly defined because the materialist has no standard by which to measure or define perfection.

The only halfway-house between the religious and the materialistic views is that of the Puritan who regards this world as but the stepping stone to another. Like so many ants our trials and tribulations are dispassionately regarded by an inscrutable deity who may, if our conduct pleases him, grant us first-class accommodation in the next world. A belief in life after death has always existed and it is irrelevant here to argue for or against it. What is certain, however, is that an absorption with such a prospect to the *exclusion* of this present life with all its infinite possibilities is disastrous, and has an egotistic effect which is little different from that of the purely materialistic view.

So chaotic and desolate is the modernist's view of this world that one of our most popular modern philosophers, arguing the case for and against the religious view, comes to the conclusion that if there is a supreme intelligence then there *must* be an after life. "If the religious view of the universe is true," he concludes, "If, that is to say, the universe has a meaning and a purpose, this life is not all, and something probably survives the break up of our bodies. Indeed, unless there is a more abundant life before mankind, this world of material things is a bad joke beyond our understanding, a vulgar laugh braying across the mysteries."[1] This is the pass to which "progress" has brought us. Were we not so blind we might perceive that we and we alone have made this world a "bad joke", and that the promise of a more abundant life exists here and not hereafter if we choose to pursue it.

In this consideration of fundamentals we may seem to have strayed far from the simple song of the Wiltshire sowers, yet this is not the case, for in the spirit which this song voices lies the key to the whole matter. For it reveals that the religious view of life, the faith and perception which is the foundation of true religion, is, like art, an acceptance of life not an escape. In its expression it is essentially humble and yet joyous, an attempt to declare absolute truth in language of symbol and image. Thus the inspiration of religion and art are one. Just as we cannot grasp the fact that holidays and holy days were once synonymous, so we cannot associate joy with religion. We inherit the barren and puritanical moralism of the nineteenth century. Once this sower's song might well have been sung in the village church to the accompaniment of the village string band without seeming incongruous. Imagine, by contrast, a church congregation singing a modern popular song with the local dance band in attendance, and we at once perceive the reason for the puritanism and sterility of modern religion, and the tragedy of modernism.

It is significant that all through the ages it has been the countryman, the man most closely in touch with natural forces and processes, who has professed the religious view of a purposive and sentient universe. If the materialist's theory were true and man's life merely an endless struggle to "conquer" blind or potentially hostile natural forces, then we might safely assume that it would gain most credence among those most closely in contact with those forces. On the contrary, we find that materialism is a philosophy of cities where man's life is largely governed by

[1] *God and Evil,* by C. E. M. Joad (1943).

forces of his own devising. What conclusions can we draw from this?

If we accept the religious, the purposive, view of the universe, we perceive its unalterable truth, not with our reason alone, for reason can deal only with material which is subject to continual change in the flux of time, but with our moral and æsthetic sense which is above reason. The materialist may choose to reject the promptings of this sense, but he can scarcely deny that it exists, nor can he afford contemptuously to dismiss its assumptions when the whole fabric of his reasoning is based upon opposite premises which, in the last analysis, are even less credible.

Wisdom is the application to life and thought of this conception of truth, and the key-note of this wisdom is the idea, or sense, of harmony and universal order. We see the universe, that is all the material and spiritual world of which we are cognizant, not as an eternal ferment of conflicting forces. but as an harmonious association of opposites in perfect balance.

The modern scientific "specialist" rejects such a view as mere mysticism, using the word in that derisive sense which he applies to any school of thought which threatens his omniscient monopoly of magic. Yet there are signs, vague as yet, but nevertheless significant and promising, that this monopoly may soon be broken, and that the scientist of the future may join hands with the artist and the visionary by gathering together the broken threads of knowledge from a past when the words "science" and "philosophy" were synonyms. It is in the field of ecology, in the study, that is, of the relationship of living things, including man himself, to each other and to their environment, that the most hopeful advances in this direction are being made. For it has been shown that life is not simply an unending struggle for the survival of the fittest, but that this "struggle" is a process by which any organisms which threaten to upset the natural order and harmony of the whole are controlled or eliminated. It is therefore only one aspect of a delicately ordered system of mutual dependencies so finely poised, so intricate, and so infinite in its scope that it is doubtful whether the human mind can ever hope to encompass a tithe of it. Remembering that man, while he has the ability to perceive this system, is, nevertheless, himself a part of it, we are naturally led to enquire what that part should be, and to conclude that it must be one of active co-operation.

Those who are now pursuing this line of research and reasoning are beginning to discover what happens when, instead of the course of co-operation, the materialist attempts to exploit and "conquer" nature in the pursuit of power. For a time such conquest may appear to succeed though the struggle becomes increasingly desperate as he advances. Despite every new weapon that man in his frenzy can invent, slowly but with inexorable certainty the legions of the flouted order multiply; the microbe, the bacillus, the fungus and the parasite. Behind them, if man does not learn his lesson, lies the desert.

If we recognize the existence of order and harmony in the universe and try to shape our lives accordingly, if, in other words, we lead an organic life in accordance with natural law, we achieve true wealth and contentment because we are fulfilling our appointed part in that harmony. This organic life, as we have seen, means the freedom to express our creative instinct, a freedom we can only realize through self-sufficiency. But because our unique creative power necessarily involves the exercise of free will we must make this way of living a deliberate aim. Were this not so, neither the effort nor the result would have any value, and we could achieve nothing. Because this life of harmony depends upon our deliberation, it follows that, alone of all created things, we can also create discord. The greater our knowledge and command of natural processes the greater the risk that, in the arrogance of this knowledge, our creative instinct will be perverted into the will-to-power, and that that ability to create harmony which is wisdom will be lost. When this occurs, as it has occurred in the modern world, we can create and can perceive only discord. We are as men caught up in a whirlwind who do not know that there is a calm at the vortex. Because our lives are discordant, separatist, and predatory, our senses can only see the same characteristics in the universe, and, like Lear, we bow beneath the weight of our own folly. In the fury of the storm we have conjured up, we either cling to the hope of survival after death or, stubborn in our materialism, continue to ride the waves in our crazy cockle shell called "Progress" in the pathetic belief that instead of being at the mercy of the elements we shall eventually steer it towards some long looked for landfall.

It has been the purpose of the previous chapters of this book to indicate that even if "Progress" succeeds in riding the storm and does not founder in irrevocable chaos as seems more probable, its landfall when sighted will prove to be no green and pleasant land of human perfectibility, but an abomination of desolation. Yet the prospect before mankind is not wholly dark. If our life is only "a vulgar laugh braying across the mysteries" it is because we have made it so, and that which we have made we may also undo. The potential for good outweighs the evil in the individual nature, and will continue to do so throughout life if it is encouraged and given freedom. The "Kingdom of God on earth" is not a misty theological chimera, it is a practicable goal within the range of our conscious ability to attain. Even to approach such an attainment may mean the labour of centuries beset with difficulty, disappointment and suffering, but were this not so the end in view would have no value. Conversely, such a common purpose justifies these means, whereas the magnitude of human suffering in the world to-day is infinitely more terrible, because it is blind and purposeless, and therefore without hope. In our agony we cannot even cry with Othello "it is the cause my soul."

The cause which we are free to pursue is that of the organic life of individual responsibility and self-sufficiency; a life in which man can play his appointed harmonic part in a greater and eternal harmony where uniformity informs diversity and where the whole exists in the part. Here mortal man, by contributing to that harmony, can ensure his immortality by becoming a part of that ageless pattern of tradition and beauty which is truth. In such a life, knowledge and wisdom, work and leisure, art and religion, life and death become indivisible parts of one art, the art of the good life. If ever, in some far distant time such a life is won, and the memory of our age seems but a dim barbaric legend, then men will understand as we can never do the innermost meaning of Stephen Harding's immortal phrase: *"Laborare est Orare."*

"CRESSY"

1942—1944.

CONCLUSIONS

For the convenience of the general reader I have summarized the argument of this book under the following twenty heads:

(1) That mediæval society was based upon a conception of man as an imperfect part of a Natural Order which demonstrated the design and purpose of a Creator.

(2) That this conception was interpreted by mediæval philosophers in the Jus Naturale, a body of Moral Law, the basis of which was that all men, of whatsoever degree, enjoyed certain rights upon condition of the acceptance of responsibility commensurate with such rights.

(3) That owing to the growth of wealth and knowledge, and to the secularization of the church, the attitude of man toward the natural world changed at the Reformation, and, in consequence, the conception upon which the Jus Naturale had been developed appeared no longer valid.

(4) That, in consequence, the concept of natural order gave place to the view of the natural world as a testing ground of vanity and temptation through which man must fight his way in the hope of eventual redemption hereafter.

(5) That this view of nature as the source of potential evil in turn prepared the way for the scientific concept of the natural world as mechanism.

(6) That the consequent decay of Moral Law has resulted in a growing insistence upon the inviolability of individual rights accompanied by the progressive abrogation of those responsibilities which such rights should involve.

(7) That this false idea of freedom by irresponsible right was, in fact, the freedom of the will-to-power, and that as such it resulted in the freedom of the few at the expense of the many.

(8) That the acquisitive society which has resulted, and the "immutable" laws by which it is believed to be governed both reflect and colour the scientific view of nature as mechanism.

(9) That a society so devoted to the pursuit of money as power necessarily frustrates the creative instinct of man by which alone he can exercise his true function as a part of the natural order.

(10) That such frustration leads to the perversion of the creative instinct either into the sadism of the will-to-power, or into the masochism of power-worship.

(11) That the disease of modern society which is leading it to chaos and barbarism cannot be healed by collectivist totalitarian methods which destroy both individual rights and responsibilities, but that regeneration depends on the recognition and restoration of both in accordance with Moral Law.

(12) That the wastage of human and natural resources which an acquisitive society necessarily incurs leads logically toward barbarism and the exhaustion of those resources.

(13) That the first step towards the arrest of this drift should be the application of the principles of natural law to education.

(14) That the conflict developing between a generation so educated and conditions of life and work becoming increasingly stultifying may lay the foundation of a renaissance.

(15) That the aim of such a renaissance would be the creation of real wealth by making the fullest use of human ability and natural resources, an aim which implies a self-sufficient society.

(16) That modern technical knowledge can be applied to this task to perform three functions, firstly to eliminate uncreative work, secondly to make the fullest use of local resources, and thirdly to provide facilities for a free and

just world trade in surpluses to overcome natural inequalities.

(17) That a self-sufficient society must necessarily be based on a prosperous and populous agricultural community, and that the purpose of industry must be to serve such a community and not vice versa.

(18) That responsible local self government is only practicable in a self-sufficient, and therefore responsible, society.

(19) That a self-sufficient society is based on the principles of art, and is therefore the only foundation of artistic renaissance.

(20) That a self-sufficient society would, through the realism of its agricultural basis, re-affirm the validity of a natural order, recover the spiritual values of religion, and perceive with reason the necessity of Moral Law.